SECOND EDITION

R Packages
Organize, Test, Document, and Share Your Code

Hadley Wickham and Jennifer Bryan

Beijing · Boston · Farnham · Sebastopol · Tokyo

R Packages

by Hadley Wickham and Jennifer Bryan

Copyright © 2023 Hadley Wickham and Jennifer Bryan. All rights reserved.

Published by O'Reilly Media, Inc., 1005 Gravenstein Highway North, Sebastopol, CA 95472.

O'Reilly books may be purchased for educational, business, or sales promotional use. Online editions are also available for most titles (*http://oreilly.com*). For more information, contact our corporate/institutional sales department: 800-998-9938 or *corporate@oreilly.com*.

Acquisitions Editor: Aaron Black	**Indexer:** Potomac Indexing, LLC
Development Editor: Sara Hunter	**Interior Designer:** David Futato
Production Editor: Aleeya Rahman	**Cover Designer:** Karen Montgomery
Copyeditor: Kim Cofer	**Illustrator:** Kate Dullea
Proofreader: Piper Editorial Consulting, LLC	

June 2023:   Second Edition

Revision History for the Second Edition

2023-06-14:   First Release

See *http://oreilly.com/catalog/errata.csp?isbn=9781098134945* for release details.

978-1-098-13494-5

[LSI]

Table of Contents

Part III.   Package Metadata

Part IV.   Testing

Part VI.   Maintenance and Distribution

Preface

Welcome!

Welcome to *R Packages* by Hadley Wickham (*https://hadley.nz*) and Jennifer Bryan (*https://jennybryan.org*). Packages are the fundamental units of reproducible R code. They include reusable R functions, the documentation that describes how to use them, and sample data. In this book you'll learn how to turn your code into packages that others can easily download and use. Writing a package can seem overwhelming at first, so start with the basics and improve it over time. It doesn't matter if your first version isn't perfect as long as the next version is better.

If you're familiar with the first edition of the book, this preface describes the major changes so that you can focus your reading on the new areas.

There are several main goals for this edition:

- Update to reflect changes in the devtools package, specifically, its "conscious uncoupling" (*https://oreil.ly/IH1YO*) into a set of smaller, more focused packages.
- Expand coverage of workflow and process, alongside the presentation of all the important moving parts that make up an R package.
- Cover entirely new topics, such as package websites and GitHub Actions (GHA).

All content has been completely revised and updated. Many chapters are new or reorganized and a couple have been removed:

- New Chapter 1, "The Whole Game" previews the entire package development process.
- New Chapter 2, "System Setup" has been carved out of the previous Introduction and gained more detail.

- The chapter formerly known as "Package Structure" has been expanded and split into two chapters, one covering package structure and state (Chapter 3) and another on workflows and tooling (Chapter 4).

- New Chapter 5, "The Package Within" demonstrates how to extract reusable logic out of data analysis scripts and into a package.

- The sections "Organizing Your Functions" and "Code Style," from Chapter 6, "R Code" have been removed, in favor of an online style guide (*https://style.tidy verse.org*). The style guide is paired with the new styler package,[1] which can automatically apply many of the rules.

- The coverage of testing has expanded into three chapters: Chapter 13 for testing basics, Chapter 14 for test suite design, and Chapter 15 for various advanced topics.

- Material around the *NAMESPACE* file and dependency relationships has been re-organized into two chapters: Chapter 10 provides technical context for thinking about dependencies, and Chapter 11 gives practice instructions for using different types of dependencies in different settings.

- New Chapter 12, "Licensing" expands earlier content on licensing into its own chapter.

- The chapter on C/C++ has been removed. It didn't have quite enough information to be useful, and since the first edition of the book, other materials have arisen that are better learning resources.

- The chapter on Git/GitHub has been reframed around the more general topic of software development practices (Chapter 20). This no longer includes step-by-step instructions for basic tasks. The use of Git/GitHub has exploded since the first edition, accompanied by an explosion of learning resources, both general and specific to R (e.g., the website Happy Git and GitHub for the useR (*https://happygitwithr.com*)). Git/GitHub still feature prominently throughout the book, most especially in Chapter 20.

- The very short *inst* chapter has been combined into Chapter 8, with all the other directories that can be important in specific contexts, but that aren't mission critical to all packages.

---

1 Kirill Müller and Lorenz Walthert, "Styler: Non-Invasive Pretty Printing of R Code," 2018. *http://styler.r-lib.org*.

Introduction

In R, the fundamental unit of shareable code is the package. A package bundles together code, data, documentation, and tests and is easy to share with others. As of March 2023, there were over 19,000 packages available on the Comprehensive R Archive Network, or CRAN, the public clearinghouse for R packages. This huge variety of packages is one of the reasons that R is so successful: the chances are that someone has already solved a problem you're working on, and you can benefit from their work by downloading their package.

If you're reading this book, you already know how to work with packages in the following ways:

- You install them from CRAN with `install.packages("x")`.
- You use them in R with `library("x")` or `library(x)`.
- You get help on them with `package?x` and `help(package = "x")`.

The goal of this book is to teach you how to develop packages so that you can write your own, not just use other people's. Why write a package? One compelling reason is that you have code that you want to share with others. Bundling your code into a package makes it easy for other people to use it, because like you, they already know how to use packages. If your code is in a package, any R user can easily download it, install it, and learn how to use it.

But packages are useful even if you never share your code. As Hilary Parker says in her introduction to packages (*https://oreil.ly/2oXTp*): "Seriously, it doesn't have to be about sharing your code (although that is an added benefit!). It is about saving yourself time." Organizing code in a package makes your life easier because packages come with conventions. For example, you put R code in *R/*, you put tests in *tests/*, and you put data in *data/*. These conventions are helpful because:

- They save time—you don't need to think about the best way to organize a project, you can just follow a template.
- Standardized conventions lead to standardized tools—if you buy into R's package conventions, you get many tools for free.

It's even possible to use packages to structure your data analyses (e.g., "Packaging Data Analytical Work Reproducibly Using r (and Friends)" in *The American Statistician* or *PeerJ Preprints*),[2] although we won't delve deeply into that use case here.

Philosophy

This book espouses our philosophy of package development: anything that can be automated, should be automated. Do as little as possible by hand. Do as much as possible with functions. The goal is to spend your time thinking about what you want your package to do rather than thinking about the minutiae of package structure.

This philosophy is realized primarily through the devtools package, which is the public face for a suite of R functions that automate common development tasks. The release of version 2.0.0 in October 2018 marked its internal restructuring into a set of more focused packages, with devtools becoming more of a metapackage. The usethis package is the subpackage you are most likely to interact with directly; we explain the devtools-usethis relationship in "devtools, usethis, and You" on page 27.

As always, the goal of devtools is to make package development as painless as possible. It encapsulates the best practices developed by Hadley Wickham, initially from his years as a prolific solo developer. More recently, he has assembled a team of developers at Posit (formerly known as RStudio), who collectively look after hundreds of open source R packages, including those known as the tidyverse (*https://www.tidyverse.org*). The reach of this team allows us to explore the space of all possible mistakes at an extraordinary scale. Fortunately, it also affords us the opportunity to reflect on both the successes and failures, in the company of expert and sympathetic colleagues. We try to develop practices that make life more enjoyable for both the maintainer and users of a package. The devtools metapackage is where these lessons are made concrete.

devtools works hand-in-hand with RStudio, which we believe is the best development environment for most R users. The most popular alternative to RStudio is currently Visual Studio Code (*https://code.visualstudio.com*) (VS Code) with the R extension (*https://oreil.ly/2f2_D*) enabled. This can be a rewarding and powerful environment; however, it does require a bit more work to set up and customize.[3]

---

2 Ben Marwick, Carl Boettiger, and Lincoln Mullen, "Packaging Data Analytical Work Reproducibly Using r (and Friends)," *The American Statistician* 72, no. 1 (2018): 80–88, *https://doi.org/10.1080/00031305.2017.1375986*; Ben Marwick, Carl Boettiger, and Lincoln Mullen, "Packaging Data Analytical Work Reproducibly Using r (and Friends)", *PeerJ Preprints* 6 (2018):e3192v2, *https://doi.org/10.7287/peerj.preprints.3192v2*.

3 Users of Emacs Speaks Statistics (*https://ess.r-project.org/*) (ESS) will find that many of the workflows described in this book are also available there. For those loyal to vim, we recommend the Nvim-R plugin (*https://github.com/jalvesaq/Nvim-R*).

RStudio

Throughout the book, we highlight specific ways that RStudio can expedite your package development workflow, in specially formatted sections like this.

Together, devtools and RStudio insulate you from the low-level details of how packages are built. As you start to develop more packages, we highly recommend that you learn more about those details. The best resource for the official details of package development is always the official *Writing R Extensions* manual (*https://oreil.ly/ Dg7M6*).[4] However, this manual can be hard to understand if you're not already familiar with the basics of packages. It's also exhaustive, covering every possible package component, rather than focusing on the most common and useful components, as this book does. *Writing R Extensions* is a useful resource once you've mastered the basics and want to learn what's going on under the hood.

In This Book

The first part of the book is all about giving you the tools you need to start your package development journey, and we highly recommend that you read it in order. We begin in Chapter 1 with a run-through of the complete development of a small package. It's meant to paint the big picture and suggest a workflow, before we descend into the detailed treatment of the key components of an R package. Then in Chapter 2 you'll learn how to prepare your system for package development, and in Chapter 3 you'll learn the basic structure of a package and how that varies across different states. Next, in Chapter 4, we'll cover the core workflows that come up repeatedly for package developers. The first part of the book ends with another case study (Chapter 5), this time focusing on how you might convert a script to a package and discussing the challenges you'll face along the way.

The remainder of the book is designed to be read as needed. Pick and choose between the chapters as the various topics come up in your development process.

First we cover key package components: Chapter 6 discusses where your code lives and how to organize it, Chapter 7 shows you how to include data in your package, and Chapter 8 covers a few less important files and directories that need to be discussed somewhere.

Next we'll dive into the package metadata, starting with *DESCRIPTION* in Chapter 9. We'll then go deep into dependencies. In Chapter 10, we'll cover the costs and benefits of taking on dependencies and provide some technical background on package namespaces and the search path. In Chapter 11, we focus on practical matters, such as

---

4 You might also enjoy the "quarto-ized" version at *https://rstudio.github.io/r-manuals/r-exts/*.

how to use different types of dependencies in different parts of your package. This is also where we discuss exporting functions, which is what makes it possible for other packages and projects to depend on your package. We'll finish off this part with a look at licensing in Chapter 12.

To ensure your package works as designed (and continues to work as you make changes), it's essential to test your code, so the next three chapters cover the art and science of testing. Chapter 13 gets you started with the basics of testing with the testthat package. Chapter 14 teaches you how to design and organize tests in the most effective way. Then we finish off our coverage of testing in Chapter 15, which teaches you advanced skills to tackle challenging situations.

If you want other people (including future-you!) to understand how to use the functions in your package, you'll need to document them. Chapter 16 gets you started using roxygen2 to document the functions in your package. Function documentation is helpful only if you know what function to look up, so next in Chapter 17 we'll discuss vignettes, which help you document the package as a whole. We'll finish up documentation with a discussion of other important markdown files like *README.md* and *NEWS.md* in Chapter 18, and creating a package website with pkgdown in Chapter 19.

The book concludes by zooming back out to consider development practices, such as the benefit of using version control and continuous integration (Chapter 20). We wrap things up by discussing the lifecycle (Chapter 21) of a package, including releasing it on CRAN (Chapter 22).

This is a lot to learn, but don't feel overwhelmed. Start with a minimal subset of useful features (e.g., just an *R/* directory!) and build up over time. To paraphrase the Zen monk Shunryu Suzuki: "Each package is perfect the way it is—and it can use a little improvement."

What's Not Here

There are also specific practices that have little to no treatment here simply because we do not use them enough to have any special insight. Does this mean that we actively discourage those practices? Probably not, as we try to be explicit about practices we think you should avoid. So if something is not covered here, it just means that a couple hundred heavily used R packages are built without meaningful reliance on that technique. That observation should motivate you to evaluate how likely it is that your development requirements truly don't overlap with ours. But sometimes the answer is a clear "yes," in which case you'll simply need to consult another resource.

Conventions Used in This Book

Throughout this book, we write `fun()` to refer to functions, `var` to refer to variables and function arguments, and *path/* for paths. Larger code blocks intermingle input and output. Output is commented so that if you have an electronic version of the book, e.g., *https://r-pkgs.org*, you can easily copy and paste examples into R. Output comments look like `#>` to distinguish them from regular comments.

The following typographical conventions are used in this book:

Italic
> Indicates new terms, URLs, email addresses, filenames, and file extensions.

`Constant width`
> Used for program listings, as well as within paragraphs to refer to program elements such as variable or function names, databases, data types, environment variables, statements, and keywords.

`Constant width bold`
> Shows commands or other text that should be typed literally by the user.

`Constant width italic`
> Shows text that should be replaced with user-supplied values or by values determined by context.

This element signifies a tip or suggestion.

This element signifies a general note.

This element indicates a warning or caution.

Colophon

This book was authored using Quarto (*https://quarto.org*) inside RStudio (*https://posit.co/products/open-source/rstudio*). The website (*https://r-pkgs.org*) is hosted with Netlify (*https://www.netlify.com*) and automatically updated after every commit by GitHub Actions. The complete source is available from GitHub (*https://github.com/hadley/r-pkgs*). This version of the book was built with:

```
library(devtools)
#> Loading required package: usethis
library(roxygen2)
library(testthat)
#>
#> Attaching package: 'testthat'
#> The following object is masked from 'package:devtools':
#>
#>     test_file

devtools::session_info()
#> ─ Session info ─────────────────────────────────────
#>  setting  value
#>  version  R version 4.2.2 (2022-10-31)
#>  os       macOS Big Sur ... 10.16
#>  system   x86_64, darwin17.0
#>  ui       X11
#>  language (EN)
#>  collate  en_US.UTF-8
#>  ctype    en_US.UTF-8
#>  tz       America/Vancouver
#>  date     2023-06-06
#>  pandoc   2.19.2 @ /Applications/RStudio.app/.../bin/tools/ (via rmarkdown)
#>
#> ─ Packages ─────────────────────────────────────────
#>  package     * version date (UTC) lib source
#>  brio          1.1.3   2021-11-30 [1] CRAN (R 4.2.0)
#>  cachem        1.0.8   2023-05-01 [1] CRAN (R 4.2.0)
#>  callr         3.7.3   2022-11-02 [1] CRAN (R 4.2.0)
#>  cli           3.6.1   2023-03-23 [1] CRAN (R 4.2.0)
#>  crayon        1.5.2   2022-09-29 [1] CRAN (R 4.2.0)
#>  devtools    * 2.4.5   2022-10-11 [1] CRAN (R 4.2.0)
#>  digest        0.6.31  2022-12-11 [1] CRAN (R 4.2.0)
#>  ellipsis      0.3.2   2021-04-29 [1] CRAN (R 4.2.0)
#>  evaluate      0.21    2023-05-05 [1] CRAN (R 4.2.0)
#>  fastmap       1.1.1   2023-02-24 [1] CRAN (R 4.2.0)
#>  fs            1.6.2   2023-04-25 [1] CRAN (R 4.2.0)
#>  glue          1.6.2   2022-02-24 [1] CRAN (R 4.2.0)
#>  htmltools     0.5.5   2023-03-23 [1] CRAN (R 4.2.2)
#>  htmlwidgets   1.6.2   2023-03-17 [1] CRAN (R 4.2.0)
#>  httpuv        1.6.9   2023-02-14 [1] CRAN (R 4.2.0)
#>  knitr         1.43    2023-05-25 [1] CRAN (R 4.2.0)
#>  later         1.3.0   2021-08-18 [1] CRAN (R 4.2.0)
```

```
#>   lifecycle      1.0.3    2022-10-07 [1] CRAN (R 4.2.0)
#>   magrittr       2.0.3    2022-03-30 [1] CRAN (R 4.2.0)
#>   memoise        2.0.1    2021-11-26 [1] CRAN (R 4.2.0)
#>   mime           0.12     2021-09-28 [1] CRAN (R 4.2.0)
#>   miniUI         0.1.1.1  2018-05-18 [1] CRAN (R 4.2.0)
#>   pkgbuild       1.4.0    2022-11-27 [1] CRAN (R 4.2.0)
#>   pkgload        1.3.2    2022-11-16 [1] CRAN (R 4.2.0)
#>   prettyunits    1.1.1    2020-01-24 [1] CRAN (R 4.2.0)
#>   processx       3.8.1    2023-04-18 [1] CRAN (R 4.2.0)
#>   profvis        0.3.7    2020-11-02 [1] CRAN (R 4.2.0)
#>   promises       1.2.0.1  2021-02-11 [1] CRAN (R 4.2.0)
#>   ps             1.7.5    2023-04-18 [1] CRAN (R 4.2.0)
#>   purrr          1.0.1    2023-01-10 [1] CRAN (R 4.2.0)
#>   R.cache        0.16.0   2022-07-21 [1] CRAN (R 4.2.0)
#>   R.methodsS3    1.8.2    2022-06-13 [1] CRAN (R 4.2.0)
#>   R.oo           1.25.0   2022-06-12 [1] CRAN (R 4.2.0)
#>   R.utils        2.12.2   2022-11-11 [1] CRAN (R 4.2.0)
#>   R6             2.5.1    2021-08-19 [1] CRAN (R 4.2.0)
#>   Rcpp           1.0.10   2023-01-22 [1] CRAN (R 4.2.0)
#>   remotes        2.4.2    2021-11-30 [1] CRAN (R 4.2.0)
#>   reprex         2.0.2    2022-08-17 [1] CRAN (R 4.2.0)
#>   rlang          1.1.1    2023-04-28 [1] CRAN (R 4.2.0)
#>   rmarkdown      2.22     2023-06-01 [1] CRAN (R 4.2.0)
#>   roxygen2     * 7.2.3    2022-12-08 [1] CRAN (R 4.2.0)
#>   rstudioapi     0.14     2022-08-22 [1] CRAN (R 4.2.0)
#>   sessioninfo    1.2.2    2021-12-06 [1] CRAN (R 4.2.0)
#>   shiny          1.7.4    2022-12-15 [1] CRAN (R 4.2.0)
#>   stringi        1.7.12   2023-01-11 [1] CRAN (R 4.2.0)
#>   stringr        1.5.0    2022-12-02 [1] CRAN (R 4.2.0)
#>   styler         1.10.1   2023-06-05 [1] CRAN (R 4.2.2)
#>   testthat     * 3.1.8    2023-05-04 [1] CRAN (R 4.2.0)
#>   urlchecker     1.0.1    2021-11-30 [1] CRAN (R 4.2.0)
#>   usethis      * 2.2.0    2023-06-06 [1] CRAN (R 4.2.2)
#>   vctrs          0.6.2    2023-04-19 [1] CRAN (R 4.2.0)
#>   withr          2.5.0    2022-03-03 [1] CRAN (R 4.2.0)
#>   xfun           0.39     2023-04-20 [1] CRAN (R 4.2.0)
#>   xml2           1.3.4    2023-04-27 [1] CRAN (R 4.2.0)
#>   xtable         1.8-4    2019-04-21 [1] CRAN (R 4.2.0)
#>   yaml           2.3.7    2023-01-23 [1] CRAN (R 4.2.0)
#>
#>   [1] /Users/jenny/Library/R/x86_64/4.2/library
#>   [2] /Library/Frameworks/R.framework/Versions/4.2/Resources/library
#>
#> ─────────────────────────────────────────────────────────────────
```

O'Reilly Online Learning

O'REILLY® For more than 40 years, *O'Reilly Media* has provided technology and business training, knowledge, and insight to help companies succeed.

Our unique network of experts and innovators share their knowledge and expertise through books, articles, and our online learning platform. O'Reilly's online learning platform gives you on-demand access to live training courses, in-depth learning paths, interactive coding environments, and a vast collection of text and video from O'Reilly and 200+ other publishers. For more information, visit *https://oreilly.com*.

How to Contact Us

Please address comments and questions concerning this book to the publisher:

> O'Reilly Media, Inc.
> 1005 Gravenstein Highway North
> Sebastopol, CA 95472
> 800-889-8969 (in the United States or Canada)
> 707-829-7019 (international or local)
> 707-829-0104 (fax)
> *support@oreilly.com*

We have a web page for this book, where we list errata, examples, and any additional information. You can access this page at *https://oreil.ly/oreillyr-packages-2e*.

For news and information about our books and courses, visit *https://oreilly.com*.

Find us on LinkedIn: *https://linkedin.com/company/oreilly-media*

Follow us on Twitter: *https://twitter.com/oreillymedia*

Watch us on YouTube: *https://www.youtube.com/oreilly*

Acknowledgments

Since the first edition of *R Packages* was published, the packages supporting the workflows described here have undergone extensive development. The original trio of devtools, roxygen2, and testthat has expanded to include the packages created by the "conscious uncoupling" of devtools, as described in "devtools, usethis, and You" on page 27. Most of these packages originate with Hadley Wickham (HW), because of their devtools roots. There are many other significant contributors, many of whom now serve as maintainers:

- devtools: HW, Winston Chang (*https://github.com/wch*), Jim Hester (*https://github.com/jimhester*) (maintainer, >= v1.13.5), Jennifer Bryan (*https://github.com/jennybc*) (maintainer >= v2.4.3)
- usethis: HW, Jennifer Bryan (*https://github.com/jennybc*) (maintainer >= v1.5.0), Malcolm Barrett
- roxygen2: HW (maintainer), Peter Danenburg (*https://github.com/klutometis*), Manuel Eugster (*https://github.com/mjaeugster*)
- testthat: HW (maintainer)
- desc: Gábor Csárdi (*https://github.com/gaborcsardi*) (maintainer), Kirill Müller (*https://github.com/krlmlr*), Jim Hester (*https://github.com/jimhester*)
- pkgbuild: HW, Jim Hester (*https://github.com/jimhester*), Gábor Csárdi (*https://github.com/gaborcsardi*) (maintainer >= v1.2.1)
- pkgload: HW, Jim Hester (*https://github.com/jimhester*), Winston Chang (*https://github.com/wch*), Lionel Henry (*https://github.com/lionel-*) (maintainer >= v1.2.4)
- rcmdcheck: Gábor Csárdi (*https://github.com/gaborcsardi*) (maintainer)
- remotes: HW, Jim Hester (*https://github.com/jimhester*), Gábor Csárdi (*https://github.com/gaborcsardi*) (maintainer), Winston Chang (*https://github.com/wch*), Martin Morgan (*https://github.com/mtmorgan*), Dan Tenenbaum (*https://github.com/dtenenba*)
- revdepcheck: HW, Gábor Csárdi (*https://github.com/gaborcsardi*) (maintainer)
- sessioninfo: HW, Gábor Csárdi (*https://github.com/gaborcsardi*) (maintainer), Winston Chang (*https://github.com/wch*), Robert Flight (*https://github.com/rmflight*), Kirill Müller (*https://github.com/krlmlr*), Jim Hester (*https://github.com/jimhester*)

This book was written and revised in the open (*https://github.com/hadley/r-pkgs/*) and it is truly a community effort: many people read drafts, fix typos, suggest improvements, and contribute content. Without those contributors, the book wouldn't be nearly as good as it is, and we are deeply grateful for their help. We are indebted to our colleagues at Posit, especially the tidyverse team, for being perpetually game to discuss package development practices. The book has been greatly improved by the suggestions from our fantastic team of technical reviewers: Malcolm Barrett, Laura DeCicco, Zhian Kamvar, Tom Mock, and Maëlle Salmon.

Thanks go to all contributors who submitted improvements via GitHub (in alphabetical order): @aaelony, @aaronwolen (Aaron Wolen), @ablejec (Andrej Blejec), @adamcduncan (Adam Duncan), @adessy, @adrtod (Adrien Todeschini), @aghaynes (Alan Haynes), @agrueneberg (Alexander Grueneberg), @alejandrohagan (Alejandro Hagan), @alesantuz (Ale Santuz), @alexandrehsd (Alexandre Henrique), @alexholcombe (Alex O. Holcombe), @alexpghayes (alex hayes), @alforj (Justin Alford),

@almartin82 (Andrew Martin), @aluxh (Alex Ho), @AmelZulji, @andreaphsz (Andrea Cantieni), @andrewdolman (Andrew Dolman), @andrewpbray (Andrew Bray), @AndrewsOR (John Andrews), @andycraig (Andrew Craig), @angela-li (Angela Li), @anjalisilva (Anjali Silva), @apomatix (Brad Friedman), @apreshill (Alison Presmanes Hill), @arashHaratian (Arash), @arilamstein (Ari Lamstein), @arneschillert (Arne Schillert), @arni-magnusson (Arni Magnusson), @asadow (Adam Sadowski), @ateucher (Andy Teucher), @avisser (Andy Visser), @ayormark (Adam Yormark), @azzaea (Azza Ahmed), @batpigandme (Mara Averick), @bclipp (Brian L), @beevabeeva, @behrman (Bill Behrman), @benmarwick (Ben Marwick), @BernhardKonrad (Bernhard Konrad), @bgreenwell (Brandon Greenwell), @Bisaloo (Hugo Gruson), @bklamer (Brett Klamer), @bm5tev3, @bms63 (Ben Straub), @bpbond (Ben Bond-Lamberty), @bquast (Bastiaan Quast), @Br-Johnson (Brett Johnson), @brews (Brewster Malevich), @brianrice2 (Brian Rice), @brry (Berry Boessenkool), @btruel, @calligross (Calli), @carldotac (Carl Lieberman), @carloscinelli (Carlos Cinelli), @CDCookJr, @cderv (Christophe Dervieux), @chambm (Matt Chambers), @charliejhadley (Charlie Joey Hadley), @chezou (Aki Ariga), @chsafouane (Safouane Chergui), @clente (Caio Lente), @cmarmstrong, @cooknl (CAPN), @CorradoLanera (Corrado Lanera), @craigcitro (Craig Citro), @crtahlin (Crt Ahlin), @daattali (Dean Attali), @danhalligan (Dan Halligan), @daroczig (Gergely Daróczi), @datarttu (Arttu Kosonen), @davidkane9 (David Kane), @DavisVaughan (Davis Vaughan), @deanbodenham, @dfalbel (Daniel Falbel), @dgrtwo (David Robinson), @dholstius (David Holstius), @DickStartz, @dkgaraujo (Douglas K. G. Araujo), @dlukes (David Lukes), @DOH-PXC5303 (Philip Crain), @dongzhuoer (Zhuoer Dong), @DougManuel (Doug Manuel), @dpprdan (Daniel Possenriede), @dracodoc (dracodoc), @drag05 (Dragos Bandur), @drvinceknight (Vince Knight), @dryzliang, @dyavorsky (Dan Yavorsky), @e-pet, @earino (E. Ariño de la Rubia), @echelleburns, @eeholmes (Eli Holmes), @eipi10 (Joel Schwartz), @ekbrown (Earl Brown), @EllaKaye (Ella Kaye), @EmilHvitfeldt (Emil Hvitfeldt), @eogoodwin, @erictleung (Eric Leung), @erikerhardt (Erik Erhardt), @espinielli (Enrico Spinielli), @ewan (Ewan Dunbar), @fbertran (Frederic Bertrand), @federicomarini (Federico Marini), @fenguoerbian (Chao Cheng), @fkohrt (Florian Kohrt), @florisvdh (Floris Vanderhaeghe), @floswald (Florian Oswald), @franrodalg (Francisco Rodríguez-Algarra), @franticspider (Simon Hickinbotham), @frycast (Daniel Vidali Fryer), @fsavje (Fredrik Sävje), @gajusmiknaitis, @gcpoole (Geoffrey Poole), @geanders (Brooke Anderson), @georoen (Jee Roen), @GerardTromp (Gerard Tromp), @GillesSanMartin (Gilles San Martin), @gmaubach (Georg Maubach), @gonzalezgouveia (Rafael Gonzalez Gouveia), @gregmacfarlane (Greg Macfarlane), @gregrs-uk (Greg), @grst (Gregor Sturm), @gsrohde (Scott Rohde), @guru809, @gustavdelius (Gustav W Delius), @haibin (Liu Haibin), @hanneoberman (Hanne Oberman), @harrismcgehee (Harris McGehee), @havenl (Haven Liu), @hcyvan (程一航), @hdraisma (Harmen), @hedderik (Hedderik van Rijn), @heists ((ᵡᵟᴏ°)ᴊ), @helske (Jouni Helske), @henningte (Henning Teickner), @HenrikBengtsson (Henrik Bengtsson), @heogden (Helen

Ogden), @hfrick (Hannah Frick), @Holzhauer (Sascha Holzhauer), @howardbaek (Howard Baek), @howbuildingsfail (How Buildings Fail), @hq9000 (Sergey Grechin), @hrbrmstr (boB Rudis), @iangow (Ian Gow), @iargent, @idmn (Iaroslav Domin), @ijlyttle (Ian Lyttle), @imchoyoung (Choyoung Im), @InfiniteCuriosity (Russ Conte), @ionut-stefanb (Ionut Stefan-Birdea), @Ironholds (Os Keyes), @ismayc (Chester Ismay), @isomorphisms (i), @jackwasey (Jack Wasey), @jacobbien (Jacob Bien), @jadeynryan (Jadey Ryan), @jameelalsalam (Jameel Alsalam), @jameslaird-smith (James Laird-Smith), @janzzon (Stefan Jansson), @JayCeBB, @jcainey (Joe Cainey), @jdblischak (John Blischak), @jedwards24 (James Edwards), @jemus42 (Lukas Burk), @jenniferthompson (Jennifer Thompson), @jeremycg (Jeremy Gray), @jgarthur (Joey Arthur), @jimhester (Jim Hester), @jimr1603 (James Riley), @jjesusfilho (José de Jesus Filho), @jkeirstead (James Keirstead), @jmarca (James Marca), @jmarshallnz (Jonathan Marshall), @joethorley (Joe Thorley), @johnbaums (John), @jolars (Johan Larsson), @jonthegeek (Jon Harmon), @jowalski (John Kowalski), @jpinelo (Joao Pinelo Silva), @jrdnbradford (Jordan), @jthomasmock (Tom Mock), @julianurbano (Julián Urbano), @jwpestrak, @jzadra (Jonathan Zadra), @jzhaoo (Joanna Zhao), @kaetschap (Sonja), @karthik (Karthik Ram), @KasperThystrup (Kasper Thystrup Karstensen), @KatherineCox, @katrinleinweber (Katrin Leinweber), @kbroman (Karl Broman), @kekecib (Ibrahim Kekec), @KellenBrosnahan, @kendonB (Kendon Bell), @kevinushey (Kevin Ushey), @kikapp (Kristopher Kapphahn), @KirkDSL, @KJByron (Karen J. Byron), @klmr (Konrad Rudolph), @KoderKow (Kyle Harris), @kokbent (Ben Toh), @kongdd (Dongdong Kong), @krlmlr (Kirill Müller), @kwenzig (Knut Wenzig), @kwstat (Kevin Wright), @kylelundstedt (Kyle G. Lundstedt), @lancelote (Pavel Karateev), @lbergelson (Louis Bergelson), @LechMadeyski (Lech Madeyski), @Lenostatos (Leon), @lindbrook, @lionel- (Lionel Henry), @LluisRamon (Lluís Ramon), @lorenzwalthert (Lorenz Walthert), @lwjohnst86 (Luke W Johnston), @maelle (Maëlle Salmon), @maiermarco, @maislind (David M), @majr-red (Matthew Roberts), @malcolmbarrett (Malcolm Barrett), @malexan (Alexander Matrunich), @manuelreif (Manuel Reif), @MarceloRTonon (Marcelo Tonon), @mariacuellar (Maria Cuellar), @markdly (Mark Dulhunty), @Marlin-Na (Marlin), @martin-mfg, @matanhakim (Matan Hakim), @matdoering, @matinang (Matina Angelopoulou), @mattflor (Matthias Flor), @maurolepore (Mauro Lepore), @maxheld83 (Max Held), @mayankvanani (Mayank Vanani), @mbjones (Matt Jones), @mccarthy-m-g (Michael McCarthy), @mdequeljoe (Matthew de Queljoe), @mdsumner (Michael Sumner), @michaelboerman (Michael Boerman), @Michael Chirico (Michael Chirico), @michaelmikebuckley (Michael Buckley), @michaelweylandt (Michael Weylandt), @miguelmorin, @MikeJohnPage, @mikelnrd (Michael Leonard), @mikelove (Mike Love), @mikemc (Michael McLaren), @MilesMcBain (Miles McBain), @mjkanji (Muhammad Jarir Kanji), @mkuehn10 (Michael Kuehn), @mllg (Michel Lang), @mohamed-180 (Mohamed El-Desokey), @moodymudskipper (Antoine Fabri), @Moohan (James McMahon), @MrAE (Jesse Leigh Patsolic), @mrcaseb, @ms609 (Martin R. Smith), @mskyttner (Markus Skyttner), @MWilson92

(Matthew Wilson), @myoung3, @nachti (Gerhard Nachtmann), @nanxstats (Nan Xiao), @nareal (Nelson Areal), @nattalides, @ncarchedi (Nick Carchedi), @ndphillips (Nathaniel Phillips), @nick-youngblut (Nick Youngblut), @njtierney (Nicholas Tierney), @nsheff (Nathan Sheffield), @osorensen (Øystein Sørensen), @PabRod (Pablo Rodríguez-Sánchez), @paternogbc (Gustavo Brant Paterno), @paulrougieux (Paul Rougieux), @pdwaggoner (Philip Waggoner), @pearsonca (Carl A. B. Pearson), @perryjer1 (Jeremiah), @petermeissner (Peter Meissner), @petersonR (Ryan Peterson), @petzi53 (Peter Baumgartner), @PhilipPallmann (Philip Pallmann), @philliplab (Phillip Labuschagne), @phonixor (Gerrit-Jan Schutten), @pkimes (Patrick Kimes), @pnovoa (Pavel Novoa), @ppanko (Pavel Panko), @pritesh-shrivastava (Pritesh Shrivastava), @PrzeChoj (PrzeChoj), @PursuitOfDataScience (Y. Yu), @pwaeckerle, @raerickson (Richard Erickson), @ramiromagno (Ramiro Magno), @ras44, @rbirkelbach (Robert Birkelbach), @rcorty (Robert W. Corty), @rdiaz02 (Ramon Diaz-Uriarte), @realAkhmed (Akhmed Umyarov), @reikookamoto (Reiko Okamoto), @renkun-ken (Kun Ren), @retowyss (Reto Wyss), @revodavid (David Smith), @rgknight (Ryan Knight), @rhgof (Richard), @rmar073, @rmflight (Robert M Flight), @rmsharp (R. Mark Sharp), @rnuske (Robert Nuske), @robertzk (Robert Krzyzanowski), @Robinlovelace (Robin Lovelace), @robiRagan (Robi Ragan), @Robsteranium (Robin Gower), @romanzenka (Roman Zenka), @royfrancis (Roy Francis), @rpruim (Randall Pruim), @rrunner, @rsangole (Rahul), @ryanatanner (Ryan), @salim-b (Salim B), @SamEdwardes (Sam Edwardes), @SangdonLim (Sangdon Lim), @sathishsrinivasank (Sathish), @sbgraves237, @schifferl (Lucas Schiffer), @scw (Shaun Walbridge), @sdarodrigues (Sabrina Rodrigues), @sebffischer (Sebastian Fischer), @serghiou (Stylianos Serghiou), @setoyama60jp, @sfirke (Sam Firke), @shannonpileggi (Shannon Pileggi), @Shelmith-Kariuki (Shel), @SheridanLGrant (Sheridan Grant), @shntnu (Shantanu Singh), @sibusiso16 (S'busiso Mkhondwane), @simdadim (Simen Buodd), @SimonPBiggs (SPB), @simonthelwall (Simon Thelwall), @SimonYansenZhao (Simon He Zhao), @singmann (Henrik Singmann), @Skenvy (Nathan Levett), @Smudgerville (Richard M. Smith), @sn248 (Satyaprakash Nayak), @sowla (Praer (Suthira) Owlarn), @srushe (Stephen Rushe), @statnmap (Sébastien Rochette), @steenharsted (Steen Harsted), @stefaneng (Stefan Eng), @stefanherzog (Stefan Herzog), @stephen-frank (Stephen Frank), @stephenll (Stephen Lienhard), @stephenturner (Stephen Turner), @stevenprimeaux (Steven Primeaux), @stevensbr, @stewid (Stefan Widgren), @sunbeomk (Sunbeom Kwon), @superdesolator (Po Su), @syclik (Daniel Lee), @symbolrush (Adrian Stämpfli-Schmid), @taekyunk (Taekyun Kim), @talgalili (Tal Galili), @tanho63 (Tan Ho), @tbrugz (Telmo Brugnara), @thisisnic (Nic Crane), @TimHesterberg (Tim Hesterberg), @titaniumtroop (Nathan), @tjebo, @tklebel (Thomas Klebel), @tmstauss (Tanner Stauss), @tonybreyal (Tony Breyal), @tonyfischetti (Tony Fischetti), @TonyLadson (Tony Ladson), @trickytank (Rick Tankard), @TroyVan, @uribo (Shinya Uryu), @urmils, @valeonte, @vgonzenbach (Virgilio Gonzenbach), @vladpetyuk (Vlad Petyuk), @vnijs (Vincent Nijs), @vspinu (Vitalie Spinu), @wcarlsen (Willi

Carlsen), @wch (Winston Chang), @wenjie2wang (Wenjie Wang), @werkstattcodes, @wiaidp, @wibeasley (Will Beasley), @wilkinson (Sean Wilkinson), @williamlief (Lief Esbenshade), @winterschlaefer (Christof Winter), @wlamnz (William Lam), @wrathematics (Drew Schmidt), @XiangyunHuang (Xiangyun Huang), @xiaochi-liu (Xiaochi), @XiaoqiLu (Xiaoqi Lu), @xiaosongz (Xiaosong Zhang), @yihui (Yihui Xie), @ynsec37, @yonicd, @ysdgroot, @yui-knk (Yuichiro Kaneko), @Zedseayou (Calum You), @zeehio (Sergio Oller), @zekiakyol (Zeki Akyol), @zenggyu (Guangyu Zeng), @zhaoy, @zhilongjia (Zhilong), @zhixunwang, @zkamvar (Zhian N. Kamvar), @zouter (Wouter Saelens)

PART I

Getting Started

## CHAPTER 1

The Whole Game

Spoiler alert!

This chapter runs through the development of a small toy package. It's meant to paint the Big Picture and suggest a workflow, before we descend into the detailed treatment of the key components of an R package.

To keep the pace brisk, we exploit the modern conveniences in the devtools package and the RStudio IDE. In later chapters, we are more explicit about what those helpers are doing for us.

This chapter is self-contained, in that completing the exercise is not a strict requirement to continue with the rest of the book; however, we strongly suggest you follow along and create this toy package with us.

Load devtools and Friends

You can initiate your new package from any active R session. You don't need to worry about whether you're in an existing or new project. The functions we use ensure that we create a new clean project for the package.

Load the devtools package, which is the public face of a set of packages that support various aspects of package development. The most obvious of these is the usethis package, which you'll see is also being loaded:

```
library(devtools)
#> Loading required package: usethis
```

Do you have an old version of devtools? Compare your version against ours and upgrade if necessary:

```
packageVersion("devtools")
#> [1] '2.4.5'
```

Toy Package: regexcite

To help walk you through the process, we use various functions from devtools to build a small toy package from scratch, with features commonly seen in released packages:

- Functions to address a specific need, in this case helpers for work with regular expressions
- Version control and an open development process
  - This is completely optional in your work but highly recommended. You'll see how Git and GitHub help us expose all the intermediate stages of our toy package.
- Access to established workflows for installation, getting help, and checking quality
  - Documentation for individual functions via roxygen2 (*https://roxygen2.r-lib.org*).
  - Unit testing with testthat (*https://testthat.r-lib.org*).
  - Documentation for the package as a whole via an executable *README.Rmd*.

We call the package *regexcite*, and it contains a couple of functions that make common tasks with regular expressions easier. Please note that these functions are very simple and we're using them here only as a means to guide you through the package development process. If you're looking for actual helpers for work with regular expressions, there are several proper R packages that address this problem space:

- stringr (*https://stringr.tidyverse.org*) (which uses stringi)
- stringi (*https://stringi.gagolewski.com*)
- rex (*https://cran.r-project.org/package=rex*)
- rematch2 (*https://cran.r-project.org/package=rematch2*)

Again, the regexcite package itself is just a device for demonstrating a typical workflow for package development with devtools.

Preview the Finished Product

The regexcite package is tracked during its development with the Git version control system. This is purely optional, and you can certainly follow along without implementing this. A nice side benefit is that we eventually connect it to a remote repository on GitHub, which means you can see the glorious result we are working toward by visiting regexcite on GitHub (*https://github.com/jennybc/regexcite*). By inspecting the commit history (*https://github.com/jennybc/regexcite/commits/main*) and especially the diffs, you can see exactly what changes at each step of the process laid out below.

create_package()

Call `create_package()` to initialize a new package in a directory on your computer. `create_package()` will automatically create that directory if it doesn't exist yet (and that is usually the case). See "Create a Package" on page 47 for more on creating packages.

Make a deliberate choice about where to create this package on your computer. It should probably be somewhere within your home directory, alongside your other R projects. It should not be nested inside another RStudio Project, R package, or Git repo. Nor should it be in an R package library, which holds packages that have already been built and installed. The conversion of the source package we create here into an installed package is part of what devtools facilitates. Don't try to do devtools' job for it!

Once you've selected where to create this package, substitute your chosen path into a `create_package()` call like this:

```
create_package("~/path/to/regexcite")
```

For the creation of this book we have to work in a temporary directory, because the book is built noninteractively in the cloud. Behind the scenes, we're executing our own `create_package()` command, but don't be surprised if our output differs a bit from yours:

```
#> ✓ Creating '/tmp/Rtmpk6VXyE/regexcite/'
#> ✓ Setting active project to '/private/tmp/Rtmpk6VXyE/regexcite'
#> ✓ Creating 'R/'
#> ✓ Writing 'DESCRIPTION'
#> Package: regexcite
#> Title: What the Package Does (One Line, Title Case)
#> Version: 0.0.0.9000
#> Authors@R (parsed):
#>     * First Last <first.last@example.com> [aut, cre] (YOUR-ORCID-ID)
#> Description: What the package does (one paragraph).
#> License: `use_mit_license()`, `use_gpl3_license()` or friends to pick a
```

```
#>      license
#> Encoding: UTF-8
#> Roxygen: list(markdown = TRUE)
#> RoxygenNote: 7.2.3
#> ✓ Writing 'NAMESPACE'
#> ✓ Writing 'regexcite.Rproj'
#> ✓ Adding '^regexcite\\.Rproj$' to '.Rbuildignore'
#> ✓ Adding '.Rproj.user' to '.gitignore'
#> ✓ Adding '^\\.Rproj\\.user$' to '.Rbuildignore'
#> ✓ Setting active project to '<no active project>'
```

If you're working in RStudio, you should find yourself in a new instance of RStudio, opened into your new regexcite package (and RStudio Project). If you somehow need to do this manually, navigate to the directory and double-click *regexcite.Rproj*. RStudio has special handling for packages, and you should now see a Build tab in the same pane as Environment and History.

You probably need to call `library(devtools)` again, because `create_package()` has probably dropped you into a fresh R session, in your new package:

```
library(devtools)
```

What's in this new directory that is also an R package and, probably, an RStudio Project? Here's a listing (locally, you can consult your Files pane):

Path	Type
.Rbuildignore	File
.gitignore	File
DESCRIPTION	File
NAMESPACE	File
R	Directory
regexcite.Rproj	File

 RStudio

In the Files pane, go to More (gear symbol) > Show Hidden Files to toggle the visibility of hidden files (a.k.a. "dotfiles" (*https://oreil.ly/cQot3*)). A select few are visible all the time, but sometimes you want to see them all.

- *.Rbuildignore* lists files that we need to have around but that should not be included when building the R package from source. If you aren't using RStudio, `create_package()` may not create this file (nor *.gitignore*) at first, since there's no RStudio-related machinery that needs to be ignored. However, you will likely develop the need for *.Rbuildignore* at some point, regardless of what editor you are using. This is discussed in more detail in ".Rbuildignore" on page 37.

- *.Rproj.user*, if you have it, is a directory used internally by RStudio.

- *.gitignore* anticipates Git usage and tells Git to ignore some standard, behind-the-scenes files created by R and RStudio. Even if you do not plan to use Git, this is harmless.

- *DESCRIPTION* provides metadata about your package. We edit this shortly, and Chapter 9 covers the general topic of the *DESCRIPTION* file.

- *NAMESPACE* declares the functions your package exports for external use and the external functions your package imports from other packages. At this point it is empty, except for a comment declaring that this is a file you should not edit by hand.

- The *R/* directory is the "business end" of your package. It will soon contain *.R* files with function definitions.

- *regexcite.Rproj* is the file that makes this directory an RStudio Project. Even if you don't use RStudio, this file is harmless. Or you can suppress its creation with `create_package(..., rstudio = FALSE)`. More in "RStudio Projects" on page 52.

use_git()

The regexcite directory is an R source package and an RStudio Project. Now we make it also a Git repository, with `use_git()`. (By the way, `use_git()` works in any project, regardless of whether it's an R package.)

```
use_git()
#> ✓ Initialising Git repo
#> ✓ Adding '.Rhistory', '.Rdata', '.httr-oauth', '.DS_Store' to '.gitignore'
```

In an interactive session, you will be asked if you want to commit some files here, and you should accept the offer. Behind the scenes, we'll also commit those same files.

So what has changed in the package? Only the creation of a *.git* directory, which is hidden in most contexts, including the RStudio file browser. Its existence is evidence that we have indeed initialized a Git repo here:

Path	Type
.git	Directory

If you're using RStudio, it probably requested permission to relaunch itself in this Project, which you should do. You can do so manually by quitting, then relaunching RStudio by double-clicking *regexcite.Rproj*. Now, in addition to package development support, you have access to a basic Git client in the Git tab of the Environment/History/Build pane.

Click History (the clock icon in the Git pane) and, if you consented, you will see an initial commit made via use_git().

RStudio

RStudio can initialize a Git repository in any Project, even if it's not an R package, as long you've set up RStudio + Git integration. Go to *Tools > Version Control > Project Setup*. Then choose "Version control system: Git" and "initialize a new git repository for this project."

Write the First Function

A fairly common task when dealing with **strings** is the need to split a single string into many parts. The strsplit() function in base R does exactly this:

```
(x <- "alfa,bravo,charlie,delta")
#> [1] "alfa,bravo,charlie,delta"
strsplit(x, split = ",")
#> [[1]]
#> [1] "alfa"    "bravo"   "charlie" "delta"
```

Take a close look at the return value:

```
str(strsplit(x, split = ","))
#> List of 1
#>  $ : chr [1:4] "alfa" "bravo" "charlie" "delta"
```

The shape of this return value often surprises people or, at least, inconveniences them. The input is a character vector of length one and the output is a list of length one. This makes total sense in light of R's fundamental tendency toward vectorization. But sometimes it's still a bit of a bummer. Often you know that your input is morally a scalar, i.e., it's just a single string, and really want the output to be the character vector of its parts.

This leads R users to employ various methods of "unlist"-ing the result:

```
unlist(strsplit(x, split = ","))
#> [1] "alfa"    "bravo"   "charlie" "delta"

strsplit(x, split = ",")[[1]]
#> [1] "alfa"    "bravo"   "charlie" "delta"
```

The second, safer solution is the basis for the inaugural function of regexcite, strsplit1():

```
strsplit1 <- function(x, split) {
  strsplit(x, split = split)[[1]]
}
```

This book does not teach you how to write functions in R. To learn more about that take a look at the Functions chapter (*https://r4ds.hadley.nz/functions.html*) of *R for Data Science* and the Functions chapter (*https://adv-r.hadley.nz/functions.html*) of *Advanced R*.

 The name of `strsplit1()` is a nod to the very handy `paste0()`, which first appeared in R 2.15.0 in 2012. `paste0()` was created to address the extremely common use case of `paste()`-ing strings together *without* a separator. `paste0()` has been lovingly described as "statistical computing's most influential contribution of the 21st century" (*https://oreil.ly/HcL-1*).

The `strsplit1()` function was so inspiring that it's now a real function in the stringr package: `stringr::str_split_1()`!

use_r()

Where should you put the definition of `strsplit1()`? Save it in a *.R* file, in the *R/* subdirectory of your package. A reasonable starting position is to make a new *.R* file for each user-facing function in your package and name the file after the function. As you add more functions, you'll want to relax this and begin to group related functions together. We'll save the definition of `strsplit1()` in the file *R/strsplit1.R*.

The helper `use_r()` creates and/or opens a script below *R/*. It really shines in a more mature package, when navigating between *.R* files and the associated test file. But even here, it's useful to keep yourself from getting too carried away while working in `Untitled4`:

```
use_r("strsplit1")
#> • Edit 'R/strsplit1.R'
```

Put the definition of `strsplit1()` *and only the definition of strsplit1()* in *R/strsplit1.R* and save it. The file *R/strsplit1.R* should *not* contain any of the other top-level code we have recently executed, such as the definition of our practice input x, `library(devtools)`, or `use_git()`. This foreshadows an adjustment you'll need to make as you transition from writing R scripts to R packages. Packages and scripts use different mechanisms to declare their dependency on other packages and to store example or test code. We explore this further in Chapter 6.

load_all()

How do we test drive `strsplit1()`? If this were a regular R script, we might use RStudio to send the function definition to the R console and define `strsplit1()` in the global environment. Or maybe we'd call `source("R/strsplit1.R")`. For package development, however, devtools offers a more robust approach.

Call `load_all()` to make `strsplit1()` available for experimentation:

```
load_all()
#> i Loading regexcite
```

Now call `strsplit1(x)` to see how it works:

```
(x <- "alfa,bravo,charlie,delta")
#> [1] "alfa,bravo,charlie,delta"
strsplit1(x, split = ",")
#> [1] "alfa"    "bravo"   "charlie" "delta"
```

Note that `load_all()` has made the `strsplit1()` function available, although it does not exist in the global environment:

```
exists("strsplit1", where = globalenv(), inherits = FALSE)
#> [1] FALSE
```

If you see `TRUE` instead of `FALSE`, that indicates you're still using a script-oriented workflow and sourcing your functions. Here's how to get back on track:

1. Clean out the global environment and restart R.

2. Reattach devtools with `library(devtools)` and reload regexcite with `load_all()`.

3. Redefine the test input x and call `strsplit1(x, split = ",")` again. This should work!

4. Run `exists("strsplit1", where = globalenv(), inherits = FALSE)` again and you should see `FALSE`.

`load_all()` simulates the process of building, installing, and attaching the regexcite package. As your package accumulates more functions—some exported, some not, some of which call each other, some of which call functions from packages you depend on—`load_all()` gives you a much more accurate sense of how the package is developing than test driving functions defined in the global environment. `load_all()` also allows much faster iteration than actually building, installing, and attaching the package. See "Test Drive with load_all()" on page 58 for more about `load_all()`.

To review what we've done so far:

- We wrote our first function, `strsplit1()`, to split a string into a character vector (not a list containing a character vector).

- We used `load_all()` to quickly make this function available for interactive use, as if we'd built and installed regexcite and attached it via `library(regexcite)`.

RStudio

RStudio exposes `load_all()` in the Build menu, in the Build pane via More > Load All, and in keyboard shortcuts Ctrl+Shift+L (Windows & Linux) or Cmd-Shift-L (macOS).

Commit strsplit1()

If you're using Git, use your preferred method to commit the new *R/strsplit1.R* file. We do so behind the scenes here, and here's the associated diff:

```
diff --git a/R/strsplit1.R b/R/strsplit1.R
new file mode 100644
index 0000000..29efb88
--- /dev/null
+++ b/R/strsplit1.R
@@ -0,0 +1,3 @@
+strsplit1 <- function(x, split) {
+  strsplit(x, split = split)[[1]]
+}
```

From this point on, we commit after each step. Remember these commits (*https:// oreil.ly/fZ3Pw*) are available in the public repository.

check()

We have informal, empirical evidence that `strsplit1()` works. But how can we be sure that all the moving parts of the regexcite package still work? This may seem silly to check, after such a small addition, but it's good to establish the habit of checking this often.

R CMD `check`, executed in the shell, is the gold standard for checking that an R package is in full working order. `check()` is a convenient way to run this without leaving your R session.

Note that `check()` produces rather voluminous output, optimized for interactive consumption. We intercept that here and just reveal a summary. Your local `check()` output will be different:

```
check()

— R CMD check results ———————————— regexcite 0.0.0.9000 ——
Duration: 8.9s

❯ checking DESCRIPTION meta-information ... WARNING
  Non-standard license specification:
    `use_mit_license()`, `use_gpl3_license()` or friends to pick a
    license
  Standardizable: FALSE

0 errors ✓ | 1 warning ✖ | 0 notes ✓
```

It is essential to actually read the output of the check! Deal with problems early and often. It's just like incremental development of *.R* and *.Rmd* files. The longer you go between full checks that everything works, the harder it becomes to pinpoint and solve your problems.

At this point, we expect 1 warning (and 0 errors, 0 notes):

```
Non-standard license specification:
  `use_mit_license()`, `use_gpl3_license()` or friends to pick a
  license
```

We'll address that soon, by doing exactly what it says. You can learn more about check() in "check() and R CMD check" on page 60.

RStudio

RStudio exposes check() in the Build menu, in the Build pane via Check, and in keyboard shortcuts Ctrl+Shift+E (Windows & Linux) or Cmd-Shift-E (macOS).

Edit DESCRIPTION

The *DESCRIPTION* file provides metadata about your package and is covered fully in Chapter 9. This is a good time to look at regexcite's current *DESCRIPTION*. You'll see it's populated with boilerplate content, which needs to be replaced.

To add your own metadata, make these edits:

- Make yourself the author. If you don't have an ORCID, you can omit the comment = ... portion.

- Write some descriptive text in the Title and Description fields.

RStudio

Use Ctrl+. in RStudio and start typing **DESCRIPTION** to activate a helper that makes it easy to open a file for editing. In addition to a filename, your hint can be a function name. This is very handy once a package has lots of files.

When you're done, `DESCRIPTION` should look similar to this:

```
Package: regexcite
Title: Make Regular Expressions More Exciting
Version: 0.0.0.9000
Authors@R:
    person("Jane", "Doe", , "jane@example.com", role = c("aut", "cre"))
Description: Convenience functions to make some common tasks with string
    manipulation and regular expressions a bit easier.
License: `use_mit_license()`, `use_gpl3_license()` or friends to pick a
    license
Encoding: UTF-8
Roxygen: list(markdown = TRUE)
RoxygenNote: 7.1.2
```

use_mit_license()

> Pick a License, Any License. (*https://oreil.ly/MKshG*)
>
> —Jeff Atwood

We currently have a placeholder in the `License` field of `DESCRIPTION` that's deliberately invalid and suggests a resolution:

```
License: `use_mit_license()`, `use_gpl3_license()` or friends to pick a
    license
```

To configure a valid license for the package, call `use_mit_license()`:

```
use_mit_license()
#> ✓ Setting License field in DESCRIPTION to 'MIT + file LICENSE'
#> ✓ Writing 'LICENSE'
#> ✓ Writing 'LICENSE.md'
#> ✓ Adding '^LICENSE\\.md$' to '.Rbuildignore'
```

This configures the `License` field correctly for the MIT license, which promises to name the copyright holders and year in a *LICENSE* file. Open the newly created *LICENSE* file and confirm it looks something like this:

```
YEAR: 2023
COPYRIGHT HOLDER: regexcite authors
```

Like other license helpers, `use_mit_license()` also puts a copy of the full license in *LICENSE.md* and adds this file to *.Rbuildignore*. It's considered a best practice to include a full license in your package's source, such as on GitHub, but CRAN disallows the inclusion of this file in a package tarball. You can learn more about licensing in Chapter 12.

document()

Wouldn't it be nice to get help on `strsplit1()`, just like we do with other R functions? This requires that your package have a special R documentation file, *man/strsplit1.Rd*, written in an R-specific markup language that is sort of like LaTeX. Luckily we don't necessarily have to author that directly.

We write a specially formatted comment right above `strsplit1()`, in its source file, and then let a package called roxygen2 (*https://roxygen2.r-lib.org*) handle the creation of *man/strsplit1.Rd*. The motivation and mechanics of roxygen2 are covered in Chapter 16.

If you use RStudio, open *R/strsplit1.R* in the source editor and put the cursor somewhere in the `strsplit1()` function definition. Now do Code > Insert roxygen skeleton. A very special comment should appear above your function, in which each line begins with #'. RStudio inserts only a barebones template, so you will need to edit it to look something like the comment below.

If you don't use RStudio, create the comment yourself. Regardless, you should modify it to look something like this:

```
#' Split a string
#'
#' @param x A character vector with one element.
#' @param split What to split on.
#'
#' @return A character vector.
#' @export
#'
#' @examples
#' x <- "alfa,bravo,charlie,delta"
#' strsplit1(x, split = ",")
strsplit1 <- function(x, split) {
  strsplit(x, split = split)[[1]]
}
```

But we're not done yet! We still need to trigger the conversion of this new roxygen comment into *man/strsplit1.Rd* with document():

```
document()
#> i Updating regexcite documentation
#> Setting `RoxygenNote` to "7.2.3"
#> i Loading regexcite
#> Writing 'NAMESPACE'
#> Writing 'strsplit1.Rd'
```

RStudio

RStudio exposes document() in the Build menu, in the Build pane via More > Document, and in keyboard shortcuts Ctrl+Shift+D (Windows & Linux) or Cmd-Shift-D (macOS).

You should now be able to preview your help file like so:

```
?strsplit1
```

You'll see a message like "Rendering development documentation for 'strsplit1'", which reminds that you are basically previewing draft documentation. That is, this documentation is present in your package's source but is not yet present in an installed package. In fact, we haven't installed regexcite yet, but we will soon. If ?strsplit1 doesn't work for you, you may need to call load_all() first, then try again.

Note also that your package's documentation won't be properly wired up until it has been formally built and installed. This polishes off niceties like the links between help files and the creation of a package index.

NAMESPACE Changes

In addition to converting strsplit1()'s special comment into *man/strsplit1.Rd*, the call to document() updates the *NAMESPACE* file, based on @export tags found in roxygen comments. Open *NAMESPACE* for inspection. The contents should be:

```
# Generated by roxygen2: do not edit by hand
```

```
export(strsplit1)
```

The export directive in *NAMESPACE* is what makes strsplit1() available to a user after attaching regexcite via library(regexcite). Just as it is entirely possible to author *.Rd* files "by hand," you can manage *NAMESPACE* explicitly yourself. But we choose to delegate this to devtools (and roxygen2). You can learn more about dependencies and *NAMESPACE* management in Chapters 10 and 11.

check() Again

regexcite should pass R CMD check cleanly now and forever more—0 errors, 0 warnings, 0 notes:

```
check()

── R CMD check results ──────────────── regexcite 0.0.0.9000 ──
Duration: 11.7s

0 errors ✓ | 0 warnings ✓ | 0 notes ✓
```

install()

Now that we know we have a minimum viable product, let's install the regexcite package into your library via install():

```
install()

── R CMD build ───────────────────────────────────────────────
* checking for file '/private/tmp/Rtmpk6VXyE/regexcite/DESCRIPTION' ... OK
* preparing 'regexcite':
* checking DESCRIPTION meta-information ... OK
* checking for LF line-endings in source and make files and shell scripts
* checking for empty or unneeded directories
* building 'regexcite_0.0.0.9000.tar.gz'
Running /Library/Frameworks/R.framework/Resources/bin/R CMD \
  INSTALL /tmp/Rtmpk6VXyE/regexcite_0.0.0.9000.tar.gz \
  --install-tests
* installing to library '/Users/jenny/Library/R/x86_64/4.2/library'
* installing _source_ package 'regexcite' ...
** using staged installation
** R
** byte-compile and prepare package for lazy loading
** help
*** installing help indices
** building package indices
** testing if installed package can be loaded from temporary location
** testing if installed package can be loaded from final location
** testing if installed package keeps a record of temporary installation path
* DONE (regexcite)
```

RStudio

RStudio exposes similar functionality in the Build menu and in the Build pane via Install and Restart, and in keyboard shortcuts Ctrl+Shift+B (Windows & Linux) or Cmd-Shift-B (macOS).

After installation is complete, we can attach and use regexcite like any other package. Let's revisit our small example from the top. This is also a good time to restart your R session and ensure you have a clean workspace:

```
library(regexcite)

x <- "alfa,bravo,charlie,delta"
strsplit1(x, split = ",")
#> [1] "alfa"    "bravo"   "charlie" "delta"
```

Success!

use_testthat()

We've tested `strsplit1()` informally, in a single example. We can formalize this as a unit test. This means we express a concrete expectation about the correct `strsplit1()` result for a specific input.

First, we declare our intent to write unit tests and to use the testthat package for this, via `use_testthat()`:

```
use_testthat()
#> ✓ Adding 'testthat' to Suggests field in DESCRIPTION
#> ✓ Setting Config/testthat/edition field in DESCRIPTION to '3'
#> ✓ Creating 'tests/testthat/'
#> ✓ Writing 'tests/testthat.R'
#> • Call `use_test()` to initialize a basic test file and open it for editing.
```

This initializes the unit testing machinery for your package. It adds `Suggests: testthat` to `DESCRIPTION`, creates the directory *tests/testthat/*, and adds the script *tests/testthat.R*. You'll notice that testthat is probably added with a minimum version of 3.0.0 and a second `DESCRIPTION` field, `Config/testthat/edition: 3`. We'll talk more about those details in Chapter 13.

However, it's still up to *you* to write the actual tests!

The helper `use_test()` opens and/or creates a test file. You can provide the file's basename or, if you are editing the relevant source file in RStudio, it will be automatically generated. For many of you, if *R/strsplit1.R* is the active file in RStudio, you can just call `use_test()`. However, since this book is built noninteractively, we must provide the basename explicitly:

```
use_test("strsplit1")
#> ✓ Writing 'tests/testthat/test-strsplit1.R'
#> • Edit 'tests/testthat/test-strsplit1.R'
```

This creates the file *tests/testthat/test-strsplit1.R*. If it had already existed, `use_test()` would have just opened it. You will notice there is an example test in the newly created file—delete that code and replace it with this content:

```
test_that("strsplit1() splits a string", {
  expect_equal(strsplit1("a,b,c", split = ","), c("a", "b", "c"))
})
```

This tests that `strsplit1()` gives the expected result when splitting a string.

Run this test interactively, as you will when you write your own. If `test_that()` or `strsplit1()` can't be found, that suggests you probably need to call `load_all()`.

Going forward, your tests mostly will run en masse and at arm's length via `test()`:

```
test()
#> i Testing regexcite
#> ✓ | F W S  OK | Context
#>
#> ⠋ |         0 | strsplit1
#> ✓ |         1 | strsplit1
#>
#> ══ Results ═══════════════════════════════════════
#> [ FAIL 0 | WARN 0 | SKIP 0 | PASS 1 ]
```

RStudio

RStudio exposes `test()` in the Build menu, in the Build pane via More > Test package, and in keyboard shortcuts Ctrl+Shift+T (Windows & Linux) or Cmd-Shift-T (macOS).

Your tests are also run whenever you `check()` the package. In this way, you basically augment the standard checks with some of your own that are specific to your package. It is a good idea to use the covr package (*https://covr.r-lib.org*) to track what proportion of your package's source code is exercised by the tests. More details can be found in Chapter 13.

use_package()

You will inevitably want to use a function from another package in your own package. We will need to use package-specific methods for declaring the other packages we need (i.e., our dependencies) and for using these packages in ours. If you plan to submit a package to CRAN, note that this even applies to functions in packages that you think of as "always available," such as `stats::median()` or `utils::head()`.

One common dilemma when using R's regular expression functions is uncertainty about whether to request `perl = TRUE` or `perl = FALSE`. And then there are often, but not always, other arguments that alter how patterns are matched, such as `fixed`, `ignore.case`, and `invert`. It can be hard to keep track of which functions use which arguments and how the arguments interact, so many users never get to the point where they retain these details without rereading the docs.

The stringr package "provides a cohesive set of functions designed to make working with strings as easy as possible." In particular, stringr uses one regular expression system everywhere (ICU regular expressions) and uses the same interface in every function for controlling matching behaviors, such as case sensitivity. Some people find this easier to internalize and program around. Let's imagine you decide you'd rather build regexcite based on stringr (and stringi) than base R's regular expression functions.

First, declare your general intent to use some functions from the stringr namespace with use_package():

```
use_package("stringr")
#> ✓ Adding 'stringr' to Imports field in DESCRIPTION
#> • Refer to functions with `stringr::fun()`
```

This adds the stringr package to the Imports field of DESCRIPTION. And that is all it does.

Let's revisit strsplit1() to make it more stringr-like. Here's a new take on it:[1]

```
str_split_one <- function(string, pattern, n = Inf) {
  stopifnot(is.character(string), length(string) <= 1)
  if (length(string) == 1) {
    stringr::str_split(string = string, pattern = pattern, n = n)[[1]]
  } else {
    character()
  }
}
```

Notice that we:

- Rename the function to str_split_one(), to signal that is a wrapper around stringr::str_split().

- Adopt the argument names from stringr::str_split(). Now we have string and pattern (and n), instead of x and split.

- Introduce a bit of argument checking and edge case handling. This is unrelated to the switch to stringr and would be equally beneficial in the version built on strsplit().

- Use the package::function() form when calling stringr::str_split(). This specifies that we want to call the str_split() function from the stringr namespace. There is more than one way to call a function from another package, and the one we endorse here is explained fully in Chapter 11.

---

1 Recall that this example was so inspiring that it's now a real function in the stringr package: stringr::str_split_1()!

Where should we write this new function definition? If we want to keep following the convention where we name the *.R* file after the function it defines, we now need to do some fiddly file shuffling. Because this comes up fairly often in real life, we have the `rename_files()` function, which choreographs the renaming of a file in *R/* and its associated companion files below *test/*:

```
rename_files("strsplit1", "str_split_one")
#> ✔ Moving 'R/strsplit1.R' to 'R/str_split_one.R'
#> ✔ Moving 'tests/testthat/test-strsplit1.R' to
#>    'tests/testthat/test-str_split_one.R'
```

Remember: the filename work is purely aspirational. We still need to update the contents of these files!

Here are the updated contents of *R/str_split_one.R*. In addition to changing the function definition, we've also updated the roxygen header to reflect the new arguments and to include examples that show off the stringr features:

```
#' Split a string
#'
#' @param string A character vector with, at most, one element.
#' @inheritParams stringr::str_split
#'
#' @return A character vector.
#' @export
#'
#' @examples
#' x <- "alfa,bravo,charlie,delta"
#' str_split_one(x, pattern = ",")
#' str_split_one(x, pattern = ",", n = 2)
#'
#' y <- "192.168.0.1"
#' str_split_one(y, pattern = stringr::fixed("."))
str_split_one <- function(string, pattern, n = Inf) {
  stopifnot(is.character(string), length(string) <= 1)
  if (length(string) == 1) {
    stringr::str_split(string = string, pattern = pattern, n = n)[[1]]
  } else {
    character()
  }
}
```

Don't forget to also update the test file!

Here are the updated contents of *tests/testthat/test-str_split_one.R*. In addition to the change in the function's name and arguments, we've added a couple more tests:

```
test_that("str_split_one() splits a string", {
  expect_equal(str_split_one("a,b,c", ","), c("a", "b", "c"))
})

test_that("str_split_one() errors if input length > 1", {
```

```
    expect_error(str_split_one(c("a,b","c,d"), ","))
})

test_that("str_split_one() exposes features of stringr::str_split()", {
  expect_equal(str_split_one("a,b,c", ",", n = 2), c("a", "b,c"))
  expect_equal(str_split_one("a.b", stringr::fixed(".")), c("a", "b"))
})
```

Before we take the new str_split_one() out for a test drive, we need to call docu
ment(). Why? Remember that document() does two main jobs:

1. Converts our roxygen comments into proper R documentation.

2. (Re)generates NAMESPACE.

The second job is especially important here, since we will no longer export
strsplit1() and we will newly export str_split_one(). Don't be dismayed by the
warning about "Objects listed as exports, but not present in namespace:
strsplit1". That always happens when you remove something from the namespace.

```
document()
#> i Updating regexcite documentation
#> i Loading regexcite
#> Warning: Objects listed as exports, but not present in namespace:
#> • strsplit1
#> Writing 'NAMESPACE'
#> Writing 'str_split_one.Rd'
#> Deleting 'strsplit1.Rd'
```

Try out the new str_split_one() function by simulating package installation via
load_all():

```
load_all()
#> i Loading regexcite
str_split_one("a, b, c", pattern = ", ")
#> [1] "a" "b" "c"
```

use_github()

You've seen us making commits during the development process for regexcite. You
can see an indicative history at *https://github.com/jennybc/regexcite*. Our use of version control and the decision to expose the development process means you can
inspect the state of the regexcite source at each developmental stage. By looking at
so-called diffs, you can see exactly how each devtools helper function modifies the
source files that constitute the regexcite package.

How would you connect your local regexcite package and Git repository to a companion repository on GitHub? Here are three approaches:

- use_github() (*https://oreil.ly/eFKpn*) is a helper that we recommend for the long term. We won't demonstrate it here because it requires some credential setup on your end. We also don't want to tear down and rebuild the public regexcite package every time we build this book.

- Set up the GitHub repo first! It sounds counterintuitive, but the easiest way to get your work onto GitHub is to initiate there, then use RStudio to start working in a synced local copy. This approach is described in Happy Git's workflows New project, GitHub first (*https://oreil.ly/5cB80*) and Existing project, GitHub first (*https://oreil.ly/uKwRu*).

- Command-line Git can always be used to add a remote repository post hoc. This is described in the Happy Git workflow Existing project, GitHub last (*https://oreil.ly/v2JZ5*).

Any of these approaches will connect your local regexcite project to a GitHub repo, public or private, which you can push to or pull from using the Git client built into RStudio. In Chapter 20, we elaborate on why version control (e.g., Git) and, specifically, hosted version control (e.g., GitHub) is worth incorporating into your package development process.

use_readme_rmd()

Now that your package is on GitHub, the *README.md* file matters. It is the package's home page and welcome mat, at least until you decide to give it a website (see Chapter 19), add a vignette (see Chapter 17), or submit it to CRAN (see Chapter 22).

The use_readme_rmd() function initializes a basic, executable *README.Rmd* ready for you to edit:

```
use_readme_rmd()
#> ✓ Writing 'README.Rmd'
#> ✓ Adding '^README\\.Rmd$' to '.Rbuildignore'
#> • Update 'README.Rmd' to include installation instructions.
#> ✓ Writing '.git/hooks/pre-commit'
```

In addition to creating *README.Rmd*, this adds some lines to *.Rbuildignore* and creates a Git precommit hook to help you keep *README.Rmd* and *README.md* in sync.

README.Rmd already has sections that prompt you to:

- Describe the purpose of the package.
- Provide installation instructions. If a GitHub remote is detected when `use_readme_rmd()` is called, this section is prefilled with instructions on how to install from GitHub.
- Show a bit of usage.

How to populate this skeleton? Copy stuff liberally from *DESCRIPTION* and any formal and informal tests or examples you have. Anything is better than nothing. This is helpful because people probably won't install your package and comb through individual help files to figure out how to use it.

We like to write the *README* in R Markdown, so it can feature actual usage. The inclusion of live code also makes it less likely that your *README* grows stale and out-of-sync with your actual package.

To make your own edits, if RStudio has not already done so, open *README.Rmd* for editing. Make sure it shows some usage of `str_split_one()`.

Here's what the *README.Rmd* (*https://oreil.ly/cenlU*) file contains:

```
---
output: github_document
---

<!-- README.md is generated from README.Rmd. Please edit that file -->

```{r, include = FALSE}
knitr::opts_chunk$set(
  collapse = TRUE,
  comment = "#>",
  fig.path = "man/figures/README-",
  out.width = "100%"
)
```

**NOTE: This is a toy package created for expository purposes, for the second
edition of [R Packages](https://r-pkgs.org). It is not meant to actually be
useful. If you want a package for factor handling, please see
[stringr](https://stringr.tidyverse.org),
[stringi](https://stringi.gagolewski.com/),
[rex](https://cran.r-project.org/package=rex), and
[rematch2](https://cran.r-project.org/package=rematch2).**
```

```
# regexcite

<!-- badges: start -->
<!-- badges: end -->

The goal of regexcite is to make regular expressions more exciting!
It provides convenience functions to make some common tasks with string
manipulation and regular expressions a bit easier.

## Installation

You can install the development version of regexcite from
[GitHub](https://github.com/) with:

``` r
# install.packages("devtools")
devtools::install_github("jennybc/regexcite")
```

Usage

A fairly common task when dealing with strings is the need to split a single
string into many parts.
This is what `base::strplit()` and `stringr::str_split()` do.

```{r}
(x <- "alfa,bravo,charlie,delta")
strsplit(x, split = ",")
stringr::str_split(x, pattern = ",")
```

Notice how the return value is a _list_ of length one, where the first element
holds the character vector of parts. Often the shape of this output is
inconvenient, i.e. we want the un-listed version.

That's exactly what `regexcite::str_split_one()` does.

```{r}
library(regexcite)

str_split_one(x, pattern = ",")
```

Use `str_split_one()` when the input is known to be a single string.
For safety, it will error if its input has length greater than one.

`str_split_one()` is built on `stringr::str_split()`, so you can use its `n`
argument and stringr's general interface for describing the `pattern` to be
matched.

```{r}
str_split_one(x, pattern = ",", n = 2)
```
```

```
y <- "192.168.0.1"
str_split_one(y, pattern = stringr::fixed("."))
```

Don't forget to render it to make *README.md*! The precommit hook should remind you if you try to commit *README.Rmd*, but not *README.md*, and also when *README.md* appears to be out-of-date.

The very best way to render *README.Rmd* is with `build_readme()`, because it takes care to render with the most current version of your package, i.e., it installs a temporary copy from the current source:

```
build_readme()
#> i Installing regexcite in temporary library
#> i Building '/private/tmp/Rtmpk6VXyE/regexcite/README.Rmd'
```

You can see the rendered *README.md* simply by visiting regexcite on GitHub (*https://github.com/jennybc/regexcite#readme*).

Finally, don't forget to do one last commit. And push, if you're using GitHub.

The End: check() and install()

Let's run `check()` again to make sure all is still well:

```
check()

— R CMD check results ──────────── regexcite 0.0.0.9000 ──
Duration: 13.4s

0 errors ✓ | 0 warnings ✓ | 0 notes ✓
```

regexcite should have no errors, warnings, or notes. This would be a good time to rebuild and install it properly. And celebrate!

```
install()

— R CMD build ─────────────────────────────────────
* checking for file '/private/tmp/Rtmpk6VXyE/regexcite/DESCRIPTION' ... OK
* preparing 'regexcite':
* checking DESCRIPTION meta-information ... OK
* checking for LF line-endings in source and make files and shell scripts
* checking for empty or unneeded directories
* building 'regexcite_0.0.0.9000.tar.gz'
Running /Library/Frameworks/R.framework/Resources/bin/R CMD \
  INSTALL /tmp/Rtmpk6VXyE/regexcite_0.0.0.9000.tar.gz \
  --install-tests
* installing to library '/Users/jenny/Library/R/x86_64/4.2/library'
* installing _source_ package 'regexcite' ...
** using staged installation
** R
** tests
```

```
** byte-compile and prepare package for lazy loading
** help
*** installing help indices
** building package indices
** testing if installed package can be loaded from temporary location
** testing if installed package can be loaded from final location
** testing if installed package keeps a record of temporary installation path
* DONE (regexcite)
```

Feel free to visit the regexcite package (*https://github.com/jennybc/regexcite*) on GitHub, which appears exactly as developed here. The commit history reflects each individual step, so use the diffs to see the addition and modification of files as the package evolved. The rest of this book goes in greater detail for each step you've seen here and much more.

Review

This chapter is meant to give you a sense of the typical package development workflow, summarized as a diagram in Figure 1-1. Everything you see here has been touched on in this chapter, with the exception of GitHub Actions, which you will learn more about in "GitHub Actions" on page 298.

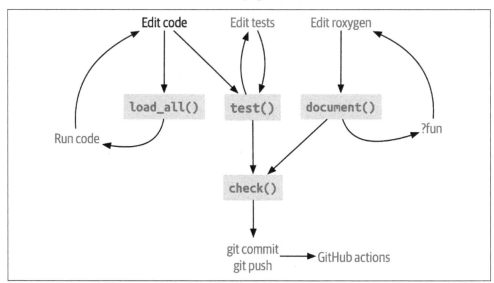

Figure 1-1. The devtools package development workflow

Here is a review of the key functions you've seen in this chapter, organized roughly by their role in the development process.

These functions set up parts of the package and are typically called once per package:

- `create_package()`
- `use_git()`
- `use_mit_license()`
- `use_testthat()`
- `use_github()`
- `use_readme_rmd()`

You will call these functions on a regular basis, as you add functions and tests or take on dependencies:

- `use_r()`
- `use_test()`
- `use_package()`

You will call these functions multiple times per day or per hour, during development:

- `load_all()`
- `document()`
- `test()`
- `check()`

## CHAPTER 2

System Setup

To get started, make sure you have the latest version of R (at least 4.2.2, which is the version being used to render this book), then run the following code to get the packages you'll need:

```
install.packages(c("devtools", "roxygen2", "testthat", "knitr"))
```

Make sure you have a recent version of the RStudio integrated development environment (IDE). New versions are released regularly, so we recommend updating often to get access to the latest and greatest features.

Download the current version of RStudio Desktop (*https://oreil.ly/94MAO*). Most readers can use the free, open source version of RStudio Desktop.

devtools, usethis, and You

> "I am large, I contain multitudes."
>
> — Walt Whitman, Song of Myself

As mentioned in "Philosophy" on page xvi, devtools is a "metapackage," encompassing and exposing functionality maintained in several smaller packages.[1] For example, devtools might provide a wrapper function in order to set user-friendly defaults, introduce helpful interactive behavior, or to combine functionality from multiple subpackages. In some cases it simply re-exports a function from another package to make it easily available when devtools is attached.

---

[1] At the time of writing, devtools exposes functionality from remotes (*https://remotes.r-lib.org*), pkgbuild (*https://pkgbuild.r-lib.org*), pkgload (*https://pkgload.r-lib.org*), rcmdcheck (*https://rcmdcheck.r-lib.org*), revdepcheck (*https://revdepcheck.r-lib.org*), sessioninfo (*https://sessioninfo.r-lib.org*), usethis (*https://usethis.r-lib.org*), testthat (*https://testthat.r-lib.org*), and roxygen2 (*https://roxygen2.r-lib.org*).

What's our recommended approach to using devtools and its constituent packages? It varies, depending on your intention:

- If you are using the functions interactively to help you develop your package, you should think of devtools as the provider of your favorite functions for package development. In this case you should attach devtools with `library(devtools)` and call the functions without qualification (e.g., `load_all()`).

- If you are using functions from devtools and friends within the package code you are writing, you should *not* depend on devtools but should instead access functions via the package that is their primary home.

  — devtools should rarely appear in the role of `pkg` in a qualified call of the form `pkg::fcn()`. Instead, `pkg` should be the package where `fcn()` is defined. For example, if you are creating a function in your package in which you need to query the state of the user's R session, use `sessioninfo::session_info()` in your package instead of `devtools::session_info()`.

- If you find bugs, try to report them on the package that is a function's primary home. The help for `devtools::fcn()` usually states when devtools is re-exporting a function from another package.

The usethis package is the one constituent package that more people may be aware of and that they may use directly. It holds the functions that act on the files and folders in an R project, most especially for any project that is also an R package. devtools makes it easy to access usethis functions interactively, as when when you call `library(devtools)`, usethis is also attached. Then you can use any function in usethis without qualification, e.g., just call `use_testthat()`. If you choose to specify the namespace, such as when working in a more programmatic style, then make sure you qualify the call with usethis, e.g., `usethis::use_testthat()`.

Personal Startup Configuration

You can attach devtools like so:

```
library(devtools)
```

But it soon grows aggravating to repeatedly attach devtools in every R session. Therefore, we strongly recommend attaching[2] devtools in your *.Rprofile* startup file, like so:

---

2 This is one of the few cases where we recommend using `require()` over `library()`. `library()` will fail with an error if it is unable to attach the package, and thus abort the execution of your *.Rprofile*. If `require()` fails to attach the package it will emit a warning but will allow the remainder of your *.Rprofile* to execute. This is discussed further in "Attaching Versus Loading" on page 151.

```
if (interactive()) {
  suppressMessages(require(devtools))
}
```

For convenience, the function use_devtools() creates .Rprofile, if needed, opens it for editing, and puts the necessary lines of code on the clipboard and the screen.

 In general, it's a bad idea to attach packages in .Rprofile, as it invites you to create R scripts that don't reflect all of their dependencies via explicit calls to library(foo). But devtools is a workflow package that smooths the process of package development and is, therefore, unlikely to get baked into any analysis scripts. Note how we still take care to attach only in interactive sessions.

usethis consults certain options when, for example, creating R packages *de novo*. This allows you to specify personal defaults for yourself as a package maintainer or for your preferred license. Here's an example of a code snippet that could go in .Rprofile:

```
options(
  usethis.description = list(
    `Authors@R` = 'person("Jane", "Doe", email = "jane@example.com")',
    License = "MIT + file LICENSE"
  )
)
```

The following code shows how to install the development versions of devtools and usethis. At times, this book may describe new features that are in the development version of devtools and related packages but that haven't been released yet:

```
devtools::install_github("r-lib/devtools")
devtools::install_github("r-lib/usethis")

# or, alternatively
pak::pak("r-lib/devtools")
pak::pak("r-lib/usethis")
```

R Build Toolchain

To be fully capable of building R packages from source, you'll also need a compiler and a few other command-line tools. This may not be strictly necessary until you want to build packages containing C or C++ code. Especially if you are using RStudio, you can set this aside for now. The IDE will alert you and provide support once you try to do something that requires you to set up your development environment. Read on for advice on doing this yourself.

Windows

On Windows the collection of tools needed for building packages from source is called Rtools.

Rtools is *not* an R package. It is *not* installed with `install.packages()`. Instead, download it from *https://cran.r-project.org/bin/windows/Rtools* and run the installer.

During the Rtools installation you may see a window asking you to "Select Additional Tasks":

- Do *not* select the box for "Edit the system PATH." devtools and RStudio should put Rtools on the `PATH` automatically when it is needed.

- Do select the box for "Save version information to registry." It should be selected by default.

macOS

You need to install the Xcode command-line tools, which requires that you register as an Apple developer (*https://oreil.ly/vQb87*) (don't worry, it's free).

Then, in the shell, do:

```
xcode-select --install
```

Alternatively, you can install the current release of full Xcode from the Mac App Store. This includes a very great deal that you do not need, but it offers the advantage of App Store convenience.

Linux

Make sure you've installed not only R but also the R development tools. For example, on Ubuntu (and Debian) you need to install the `r-base-dev` package with:

```
sudo apt install r-base-dev
```

On Fedora and RedHat, the development tools (called `R-core-devel`) will be installed automatically when you install with R with `sudo dnf install R`.

Verify System Prep

You can request a "(package) development situation report" with `devtools::dev_sitrep()`:

```
devtools::dev_sitrep()
#> ── R ─────────────────────────────────────────────────────
#> • version: 4.1.2
#> • path: '/Library/Frameworks/R.framework/Versions/4.1/Resources/'
#> ── RStudio ───────────────────────────────────────────────
```

```
#> • version: 2022.2.0.443
#> — devtools ────────────────────────────────────────
#> • version: 2.4.3.9000
#> • devtools or its dependencies out of date:
#>   'gitcreds', 'gh'
#>   Update them with `devtools::update_packages("devtools")`
#> — dev package ─────────────────────────────────────
#> • package: 'rpkgs'
#> • path: '/Users/jenny/rrr/r-pkgs/'
#> • rpkgs dependencies out of date:
#>   'gitcreds', 'generics', 'tidyselect', 'dplyr', 'tidyr', 'broom', 'gh'
#>   Update them with `devtools::install_dev_deps()`
```

If this reveals that certain tools or packages are missing or out-of-date, you are encouraged to update them.

## CHAPTER 3

Package Structure and State

This chapter will start you on the road to package development by converting the implicit knowledge you've gained from *using* R packages into the explicit knowledge needed to *create and modify* them. You'll learn about the various states a package can be in and the difference between a package and library (and why you should care).

Package States

When you create or modify a package, you work on its "source code" or "source files." You interact with the in-development package in its *source* form. This is *not* the package form you are most familiar with from day-to-day use. Package development workflows make much more sense if you understand the five states an R package can be in:

- Source
- Bundled
- Binary
- Installed
- In-memory

You already know some of the functions that put packages into these states. For example, install.packages() can move a package from the source, bundled, or binary states into the installed state. devtools::install_github() takes a source package on GitHub and moves it into the installed state. The library() function loads an installed package into memory, making it available for immediate and direct use.

Source Package

A *source* package is just a directory of files with a specific structure. It includes particular components, such as a *DESCRIPTION* file, an *R/* directory containing *.R* files, and so on. Most of the remaining chapters in this book are dedicated to detailing these components.

If you are new to package development, you may have never seen a package in source form! You might not even have any source packages on your computer. The easiest way to see a package in source form right away is to browse around its code on the web.

Many R packages are developed in the open on GitHub (or GitLab or similar). The best-case scenario is that you visit the package's CRAN landing page, e.g.:

- forcats: (*https://cran.r-project.org/package=forcats*)
- readxl: (*https://cran.r-project.org/package=readxl*)

and one of its URLs links to a repository on a public hosting service, e.g.:

- forcats: (*https://github.com/tidyverse/forcats*)
- readxl: (*https://github.com/tidyverse/readxl*)

Some maintainers forget to list this URL, even though the package is developed in a public repository, but you still might be able to discover it via search.

Even if a package is not developed on a public platform, you can visit its source in the unofficial, read-only mirror maintained by R-hub (*https://docs.r-hub.io/#cranatgh*). Examples:

- MASS: (*https://github.com/cran/MASS*)
- car: (*https://github.com/cran/car*)

Note that exploring a package's source and history within the `cran` GitHub organization is not the same as exploring the package's true development venue, because this source and its evolution is just reverse-engineered from the package's CRAN releases. This presents a redacted view of the package and its history, but, by definition, it includes everything that is essential.

Bundled Package

A *bundled* package is a package that's been compressed into a single file. By convention (from Linux), package bundles in R use the extension *.tar.gz* and are sometimes referred to as "source tarballs." This means that multiple files have been reduced to a single file (*.tar*) and then compressed using gzip (*.gz*). While a bundle is not that useful on its own, it's a platform-agnostic, transportation-friendly intermediary between a source package and an installed package.

In the rare case that you need to make a bundle from a package you're developing locally, use `devtools::build()`. Under the hood, this calls `pkgbuild::build()` and, ultimately, `R CMD build`, which is described further in the "Building package tarballs" section of *Writing R Extensions* (*https://oreil.ly/LNdL3*).

This should tip you off that a package bundle or "source tarball" is not simply the result of making a tar archive of the source files, then compressing with gzip. By convention, in the R world, a few more operations are carried out when making the *.tar.gz* file, and this is why we've elected to refer to this form as a package *bundle* in this book.

Every CRAN package is available in bundled form, via the "Package source" field of its landing page. Continuing our examples, you could download the bundles `forcats_0.4.0.tar.gz` and `readxl_1.3.1.tar.gz` (or whatever the current versions may be). You could unpack such a bundle in the shell (not the R console) like so:

```
tar xvf forcats_0.4.0.tar.gz
```

If you decompress a bundle, you'll see it looks almost the same as a source package. Figure 3-1 shows the files present in the source, bundled, and binary forms of a fictional package named zzzpackage. We've deliberately crafted this example to include most of the package parts covered in this book. Not every package will include every file seen here, nor does this diagram include every possible file that might appear in a package.

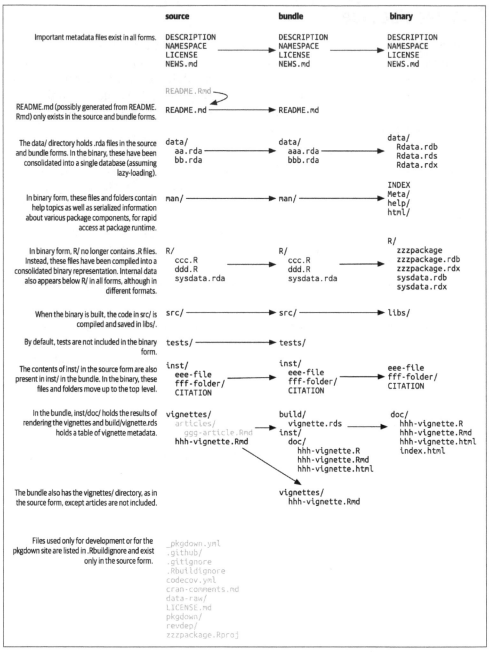

Figure 3-1. Package forms: source versus bundled versus binary

The main differences between a source package and an uncompressed bundle are:

- Vignettes have been built, so rendered outputs, such as HTML, appear below *inst/doc/* and a vignette index appears in the *build/* directory.

- A local source package might contain temporary files used to save time during development, like compilation artifacts in *src/*. These are never found in a bundle.

- Any files listed in *.Rbuildignore* are not included in the bundle. These are typically files that facilitate your development process, but they should be excluded from the distributed product.

.Rbuildignore

You won't need to contemplate the exact structure of package *.tar.gz* files very often, but you do need to understand the *.Rbuildignore* file. It controls which files from the source package make it into the downstream forms.

Each line of *.Rbuildignore* is a Perl-compatible regular expression that is matched, without regard to case, against the path to each file in the source package.[1] If the regular expression matches, that file or directory is excluded. Note there are some default exclusions implemented by R itself, mostly relating to classic version control systems and editors, such as SVN, Git, and Emacs.

We usually modify *.Rbuildignore* with the usethis::use_build_ignore() function, which takes care of easy-to-forget details, such as regular expression anchoring and escaping. To exclude a specific file or directory (the most common use case), you *must* anchor the regular expression. For example, to exclude a directory called "notes," the *.Rbuildignore* entry must be ^notes$, whereas the unanchored regular expression notes will match any filename containing "notes," e.g., *R/notes.R, man/important-notes.R, data/endnotes.Rdata*, etc. We find that use_build_ignore() helps us get more of our *.Rbuildignore* entries right the first time.

.Rbuildignore is a way to resolve some of the tension between the practices that support your development process and CRAN's requirements for submission and distribution (Chapter 22). Even if you aren't planning to release on CRAN, following these conventions will allow you to make the best use of R's built-in tooling for package checking and installation. The files you should *.Rbuildignore* fall into two broad, semi-overlapping classes:

---

1 To see the set of filepaths that should be on your radar, execute dir(full.names = TRUE, recursive = TRUE, include.dirs = TRUE, all.files = TRUE) in the package's top-level directory.

- Files that help you generate package contents programmatically. Examples:
  - Using *README.Rmd* to generate an informative and current *README.md* (see "README" on page 273)
  - Storing *.R* scripts to create and update internal or exported data (see "Preserve the Origin Story of Package Data" on page 102)
- Files that drive package development, checking, and documentation, outside of CRAN's purview. Examples:
  - Files relating to the RStudio IDE (see "RStudio Projects" on page 52)
  - Using the pkgdown package (*https://pkgdown.r-lib.org*) to generate a website (see Chapter 19)
  - Configuration files related to continuous integration/deployment (see "Continuous Integration" on page 298)

Here is a nonexhaustive list of typical entries in the *.Rbuildignore* file for a package in the tidyverse:

```
^.*\.Rproj$           # Designates the directory as an RStudio Project
^\.Rproj\.user$       # Used by RStudio for temporary files
^README\.Rmd$         # An Rmd file used to generate README.md
^LICENSE\.md$         # Full text of the license
^cran-comments\.md$   # Comments for CRAN submission
^data-raw$            # Code used to create data included in the package
^pkgdown$             # Resources used for the package website
^_pkgdown\.yml$       # Configuration info for the package website
^\.github$            # GitHub Actions workflows
```

Note that the comments must not appear in an actual *.Rbuildignore* file; they are included here only for exposition.

We'll mention when you need to add files to *.Rbuildignore* whenever it's important. Remember that `usethis::use_build_ignore()` is an attractive way to manage this file. Furthermore, many usethis functions that add a file that should be listed in *.Rbuildignore* take care of this automatically. For example, use_read_rmd() adds `^README\.Rmd$` to *.Rbuildignore*.

Binary Package

If you want to distribute your package to an R user who doesn't have package development tools, you'll need to provide a *binary* package. The primary maker and distributor of binary packages is CRAN, not individual maintainers. But even if you delegate the responsibility of distributing your package to CRAN, it's still important for a maintainer to understand the nature of a binary package.

Like a package bundle, a binary package is a single file. Unlike a bundled package, a binary package is platform specific and there are two basic flavors: Windows and macOS. (Linux users are generally required to have the tools necessary to install from *.tar.gz* files, although the emergence of resources like Posit Public Package Manager (*https://packagemanager.posit.co*) is giving Linux users the same access to binary packages as their colleagues on Windows and macOS.)

Binary packages for macOS are stored as *.tgz*, whereas Windows binary packages end in *.zip*. If you need to make a binary package, use `devtools::build(binary = TRUE)` on the relevant operating system. Under the hood, this calls `pkgbuild::build(binary = TRUE)` and, ultimately, `R CMD INSTALL --build`, which is described further in the "Building binary packages" section of *Writing R Extensions* (*https://oreil.ly/hIfxL*). If you choose to release your package on CRAN (Chapter 22), you submit your package in bundled form, then CRAN creates and distributes the package binaries.

CRAN packages are usually available in binary form, for both macOS and Windows, for the current, previous, and (possibly) development versions of R. Continuing our examples from earlier, you could download binary packages such as:

- forcats for macOS (`forcats_0.4.0.tgz`)
- readxl for Windows (`readxl_1.3.1.zip`)

and this is, indeed, part of what's usually going on behind the scenes when you call `install.packages()`.

If you uncompress a binary package, you'll see that the internal structure is rather different from a source or bundled package. Figure 3-1 includes this comparison, so this is a good time to revisit that diagram. Here are some of the most notable differences:

- There are no *.R* files in the *R/* directory; instead, there are three files that store the parsed functions in an efficient file format. This is basically the result of loading all the R code and then saving the functions with `save()`. (In the process, this adds a little extra metadata to make things as fast as possible.)
- A *Meta/* directory contains a number of *.rds* files. These files contain cached metadata about the package, like what topics the help files cover and a parsed version of the *DESCRIPTION* file. (You can use `readRDS()` to see exactly what's in those files.) These files make package loading faster by caching costly computations.
- The actual help content appears in *help/* and *html/* (no longer in *man/*).
- If you had any code in the *src/* directory, there will now be a *libs/* directory that contains the results of compiling the code.

- If you had any objects in *data/*, they have now been converted into a more efficient form.
- The contents of *inst/* are moved to the top-level directory. For example, vignette files are now in *doc/*.
- Some files and folders have been dropped, such as *README.md*, *build/*, *tests/*, and *vignettes/*.

Installed Package

An *installed* package is a binary package that's been decompressed into a package library (described in "Package Libraries" on page 43). Figure 3-2 illustrates the many ways a package can be installed, along with a few other functions for converting a package from one state to another. This diagram is complicated! In an ideal world, installing a package would involve stringing together a set of simple steps: source → bundle, bundle → binary, binary → installed. In the real world, it's not this simple because there are often (faster) shortcuts available.

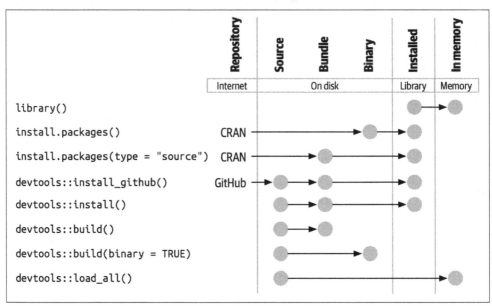

Figure 3-2. Many methods for converting between package states

The built-in command-line tool R CMD INSTALL powers all package installation. It can install a package from source files, a bundle (a.k.a. a source tarball), or a binary package. Details are available in the "Installing packages" section of *R Installation and Administration* (*https://oreil.ly/pLH_H*). Just like with devtools::build(), devtools provides a wrapper function, devtools::install(), that makes this tool available from within an R session.

RStudio

RStudio can also help you install your in-development package via the *Install* and *More* drop-downs in the *Build* pane and with *Install Package* in the *Build* menu.

Most useRs understandably like to install packages from the comfort of an R session and directly from CRAN. The built-in function `install.packages()` meets this need. It can download the package in various forms, install it, and optionally attend to the installation of dependencies.

There is a price, however, for the convenience of installing R packages from within an R session. As you might expect, it can be a bit tricky to reinstall a package that is already in use in the current session. This actually works most of the time, but sometimes it does not, especially when installing an R package with compiled code on Windows. Due to how file handles are locked on Windows, an attempt to install a new version of a package that's in use can result in a corrupt installation where the package's R code has been updated, but its compiled code has not. When troubleshooting, Windows users should strive to install packages in a clean R session, with as few packages loaded as possible.

The pak package (*https://pak.r-lib.org*) is a relative newcomer (at the time of writing) and provides a promising alternative to `install.packages()`, as well as other more specialized functions such as `devtools::install_github()`. It's too early to make a blanket recommendation for using pak for all of your package installation needs, but we are certainly using it more and more in our personal workflows. One of pak's flagship features is that it nicely solves the "locked DLL" problem just described, i.e., updating a package with compiled code on Windows. As you get deeper into package development, you will find yourself doing a whole new set of tasks, such as installing a dependency from an in-development branch or scrutinizing package dependency trees. pak provides a rich toolkit for this and many other related tasks. We predict that pak will soon become our official recommendation for how to install packages (and more).

However, in the meantime, we describe the status quo. devtools has long offered a family of `install_*()` functions to address some needs beyond the reach of `install.packages()` or to make existing capabilities easier to access. These functions are actually maintained in the remotes package (*https://remotes.r-lib.org*) and are re-exported by devtools. (Given what we said earlier, it is likely that remotes will essentially become superseded, in favor of pak, but we're not quite there yet.)

```
library(remotes)

funs <- as.character(lsf.str("package:remotes"))
grep("^install_.+", funs, value = TRUE)
#>  [1] "install_bioc"       "install_bitbucket" "install_cran"
#>  [4] "install_deps"       "install_dev"        "install_git"
#>  [7] "install_github"     "install_gitlab"     "install_local"
#> [10] "install_remote"     "install_svn"        "install_url"
#> [13] "install_version"
```

`install_github()` is the most useful of these functions and is also featured in
Figure 3-2. It is the flagship example of a family of functions that can download
a package from a remote location that is not CRAN and do whatever is necessary
to install it and its dependencies. The rest of the devtools/remotes `install_*()`
functions are aimed at making things that are technically possible with base tooling
a bit easier or more explicit, such as `install_version()`, which installs a specific
version of a CRAN package.

Analogous to *.Rbuildignore*, described in section ".Rbuildignore" on page
37, *.Rinstignore* lets you keep files present in a package bundle out of the installed
package. However, in contrast to *.Rbuildignore*, this is rather obscure and rarely
needed.

In-Memory Package

We finally arrive at a command familiar to everyone who uses R:

```
library(usethis)
```

Assuming usethis is installed, this call makes its functions available for use; i.e., now
we can do:

```
create_package("/path/to/my/coolpackage")
```

The usethis package has been loaded into memory and, in fact, has also been attached
to the search path. The distinction between loading and attaching packages is not
important when you're writing scripts, but it's very important when you're writing
packages. You'll learn more about the difference and why it's important in "Attaching
Versus Loading" on page 151.

`library()` is not a great way to iteratively tweak and test drive a package you're
developing, because it works only for an installed package. In "Test Drive with
load_all()" on page 58, you'll learn how `devtools::load_all()` accelerates development by allowing you to load a source package directly into memory.

Package Libraries

We just discussed the library() function, whose name is inspired by what it does. When you call library(somepackage), R looks through the current *libraries* for an installed package named "somepackage" and, if successful, it makes somepackage available for use.

In R, a *library* is a directory containing installed packages, sort of like a library for books. Unfortunately, in the R world, you will frequently encounter confused usage of the words "library" and "package." It's common for someone to refer to dplyr, for example, as a library when it is actually a package. There are a few reasons for the confusion. First, R's terminology arguably runs counter to broader programming conventions, where the usual meaning of "library" is closer to what we mean by "package." The name of the library() function itself probably reinforces the wrong associations. Finally, this vocabulary error is often harmless, so it's easy for R users to fall into the wrong habit and for people who point out this mistake to look like insufferable pedants. But here's the bottom line:

> We use the library() function to load [2] a *package*.

The distinction between the two is important and useful as you get involved in package development.

You can have multiple libraries on your computer. In fact, many of you already do, especially if you're on Windows. You can use .libPaths() to see which libraries are currently active. Here's how this might look on Windows:

```
# on Windows
.libPaths()
#> [1] "C:/Users/jenny/Documents/R/win-library/4.2"
#> [2] "C:/Program Files/R/R-4.2.2/library"

lapply(.libPaths(), list.dirs, recursive = FALSE, full.names = FALSE)
#> [[1]]
#>   [1] "abc"          "anytime"      "askpass"      "assertthat"
#>   ...
#> [145] "zeallot"
#>
#> [[2]]
#>   [1] "base"         "boot"         "class"        "cluster"
#>   [5] "codetools"    "compiler"     "datasets"     "foreign"
#>   [9] "graphics"     "grDevices"    "grid"         "KernSmooth"
#> [13] "lattice"      "MASS"         "Matrix"       "methods"
#> [17] "mgcv"         "nlme"         "nnet"         "parallel"
```

---

2 Well, actually, library() loads and attaches a package, but that's a topic for "Attaching Versus Loading" on page 151.

```
#> [21] "rpart"        "spatial"      "splines"  "stats"
#> [25] "stats4"       "survival"     "tcltk"    "tools"
#> [29] "translations" "utils"
```

Here's a similar look on macOS (but your results may vary):

```
# on macOS
.libPaths()
#> [1] "/Users/jenny/Library/R/arm64/4.2/library"
#> [2] "/Library/Frameworks/R.framework/Versions/4.2-arm64/Resources/library"

lapply(.libPaths(), list.dirs, recursive = FALSE, full.names = FALSE)
#> [[1]]
#>    [1] "abc"                    "abc.data"            "abind"
#>  ...
#> [1033] "Zelig"                  "zip"                 "zoo"
#>
#> [[2]]
#>  [1] "base"         "boot"         "class"     "cluster"
#>  [5] "codetools"    "compiler"     "datasets"  "foreign"
#>  [9] "graphics"     "grDevices"    "grid"      "KernSmooth"
#> [13] "lattice"      "MASS"         "Matrix"    "methods"
#> [17] "mgcv"         "nlme"         "nnet"      "parallel"
#> [21] "rpart"        "spatial"      "splines"   "stats"
#> [25] "stats4"       "survival"     "tcltk"     "tools"
#> [29] "translations" "utils"
```

In both cases we see two active libraries, consulted in this order:

1. A user library
2. A system-level or global library

This setup is typical on Windows but is something you usually need to opt into on macOS and Linux.[3] With this setup, add-on packages installed from CRAN (or elsewhere) or under local development are kept in the user library. In the preceding macOS example, the system is used as a primary development machine and has many packages here (~1000), whereas the Windows system is used only occasionally and is much more spartan. The core set of base and recommended packages that ship with R live in the system-level library and are the same on all operating systems. This separation appeals to many developers and makes it easy to, for example, clean out your add-on packages without disturbing your base R installation.

If you're on macOS or Linux and only see one library, there is no urgent need to change anything. But next time you upgrade R, consider creating a user-level

---

3 For more details, see the Maintaining R section (*https://oreil.ly/G1VuL*) in *What They Forgot To Teach You About R*, *Managing Libraries* (*https://oreil.ly/db45k*) in *R Installation and Administration*, and the R help files for ?Startup and ?.libPaths.

library. By default, R looks for a user library found at the path stored in the environment variable R_LIBS_USER, which itself defaults to ~/Library/R/m/x.y/library on macOS, and ~/R/m-library/x.y on Linux (where m is a concise description of your CPU architecture, and x.y is the R version). You can see this path with Sys.getenv("R_LIBS_USER"). These directories do not exist by default, and the use of them must be enabled by creating the directory. When you install a new version of R, and prior to installing any add-on packages, use dir.create(Sys.getenv("R_LIBS_USER"), recursive = TRUE) to create a user library in the default location. Now you will have the recommended library setup. Alternatively, you could set up a user library elsewhere and tell R about that by setting the R_LIBS_USER environment variable in *.Renviron*. The simplest way to edit your *.Renviron* file is with usethis::edit_r_environ(), which will create the file if it doesn't exist and open it for editing.

The filepaths for these libraries also make it clear they are associated with a specific version of R (4.2.x at the time of writing), which is also typical. This reflects and enforces the fact that you need to reinstall your add-on packages when you update R from, say, 4.1 to 4.2, which is a change in the *minor* version. You generally do not need to reinstall add-on packages for a *patch* release, e.g., going from R 4.2.1 to 4.2.2.

As your R usage grows more sophisticated, it's common to start managing package libraries with more intention. For example, tools like renv (*https://rstudio.github.io/renv/*) (and its predecessor packrat (*https://rstudio.github.io/packrat/*)) automate the process of managing project-specific libraries. This can be important for making data products reproducible, portable, and isolated from one another. A package developer might prepend the library search path with a temporary library, containing a set of packages at specific versions, in order to explore issues with backward and forward compatibility, without affecting other day-to-day work. Reverse dependency checks are another example where we explicitly manage the library search path.

Here are the main levers that control which libraries are active, in order of scope and persistence:

- Environment variables, like R_LIBS and R_LIBS_USER, which are consulted at startup.
- Calling .libPaths() with one or more filepaths.
- Executing small snippets of code with a temporarily altered library search path via withr::with_libpaths().
- Arguments to individual functions, like install.packages(lib =) and library(lib.loc =).

Finally, it's important to note that `library()` should *never* be used *inside a package*. Packages and scripts rely on different mechanisms for declaring their dependencies, and this is one of the biggest adjustments you need to make in your mental model and habits. We explore this topic fully in "Imports, Suggests, and Friends" on page 129 and Chapter 11.

## CHAPTER 4

Fundamental Development Workflows

Having peeked under the hood of R packages and libraries in Chapter 3, here we provide the basic workflows for creating a package and moving it through the different states that come up during development.

Create a Package

Many packages are born out of one person's frustration at some common task that should be easier. How should you decide whether something is package-worthy? There's no definitive answer, but it's helpful to appreciate at least two types of payoff:

Product
> Your life will be better when this functionality is implemented formally, in a package.

Process
> Greater mastery of R will make you more effective in your work.

Survey the Existing Landscape

If all you care about is the existence of a product, then your main goal is to navigate the space of existing packages. Silge, Nash, and Graves organized a survey and sessions around this at useR! 2017 and their write up for *The R Journal* provides a comprehensive roundup of resources.[1]

If you are looking for ways to increase your R mastery, you should still educate yourself about the landscape. But there are plenty of good reasons to make your

---

[1] Julia Silge, John C. Nash, and Spencer Graves, "Navigating the R Package Universe," *The R Journal* 10, no. 2 (2018): 558–63. *https://doi.org/10.32614/RJ-2018-058*.

own package, even if there is relevant prior work. The way experts got that way is by actually building things, often very basic things, and you deserve the same chance to learn by tinkering. If you're only allowed to work on things that have never been touched, you're likely looking at problems that are either very obscure or very difficult.

It's also valid to evaluate the suitability of existing tools on the basis of user interface, defaults, and edge-case behavior. A package may technically do what you need, but perhaps it's very unergonomic for your use case. In this case, it may make sense for you to develop your own implementation or to write wrapper functions that smooth over the sharp edges.

If your work falls into a well-defined domain, educate yourself about the existing R packages, even if you've resolved to create your own package. Do they follow specific design patterns? Are there specific data structures that are common as the primary input and output? For example, there is a very active R community around spatial data analysis (*https://www.r-spatial.org*) that has successfully self-organized to promote greater consistency across packages with different maintainers. In modeling, the hardhat package (*https://hardhat.tidymodels.org*) provides scaffolding for creating a modeling package that plays well with the tidymodels (*https://www.tidymodels.org*) ecosystem. Your package will get more usage and will need less documentation if it fits nicely into the surrounding landscape.

Name Your Package

> "There are only two hard things in Computer Science: cache invalidation and naming things."
>
> — Phil Karlton

Before you can create your package, you need to come up with a name for it. This can be the hardest part of creating a package! (Not least because no one can automate it for you.)

Formal requirements

There are three formal requirements:

- The name can only consist of letters, numbers, and periods, i.e., ..
- It must start with a letter.
- It cannot end with a period.

Unfortunately, this means you can't use either hyphens or underscores, i.e., - or _, in your package name. We recommend against using periods in package names, due to confusing associations with file extensions and S3 methods.

Things to consider

If you plan to share your package with others, it's important to come up with a good name. Here are some tips:

- Pick a unique name that's easy to Google. This makes it easy for potential users to find your package (and associated resources) and for you to see who's using it.

- Don't pick a name that's already in use on CRAN or Bioconductor. You may also want to consider some other types of name collision:
  - Is there an in-development package maturing on, say, GitHub that already has some history and seems to be heading toward release?
  - Is this name already used for another piece of software or for a library or framework in, e.g., the Python or JavaScript ecosystem?

- Avoid using both upper- and lower-case letters; doing so makes the package name hard to type and even harder to remember. For example, it's hard to remember if it's Rgtk2 or RGTK2 or RGtk2.

- Give preference to names that are pronounceable, so people are comfortable talking about your package and have a way to hear it inside their head.

- Find a word that evokes the problem and modify it so that it's unique. Here are some examples:
  - lubridate makes dates and times easier.
  - rvest "harvests" the content from web pages.
  - r2d3 provides utilities for working with D3 visualizations.
  - forcats is an anagram of factors, which we use *for* categorical data.

- Use abbreviations, like the following:
  - Rcpp = R + C++ (plus plus)
  - brms = Bayesian Regression Models using Stan

- Add an extra R, for example:
  - stringr provides string tools.
  - beepr plays notification sounds.
  - callr calls R, from R.

- Don't get sued.
  - If you're creating a package that talks to a commercial service, check the branding guidelines. For example, rDrop isn't called rDropbox because Dropbox prohibits any applications from using the full trademarked name.

Nick Tierney presents a fun typology of package names in his Naming Things blog post (*https://oreil.ly/TPrwz*), which also includes more inspiring examples. He also has some experience with renaming packages; the post "So, you've decided to change your r package name" (*https://oreil.ly/pb0Al*) is a good resource if you don't get this right the first time.

Use the available package

It is impossible to abide by all of the previous suggestions simultaneously, so you will need to make some trade-offs. The available package (*https://oreil.ly/xYO5i*) has a function called `available()` that helps you evaluate a potential package name from many angles:

```
library(available)

available("doofus")
#> Urban Dictionary can contain potentially offensive results,
#>   should they be included? [Y]es / [N]o:
#> 1: 1
#> — doofus ─────────────────────────────────────────────────────
#> Name valid: ✓
#> Available on CRAN: ✓
#> Available on Bioconductor: ✓
#> Available on GitHub:  ✓
#> Abbreviations: http://www.abbreviations.com/doofus
#> Wikipedia: https://en.wikipedia.org/wiki/doofus
#> Wiktionary: https://en.wiktionary.org/wiki/doofus
#> Sentiment:???
```

`available::available()` does the following:

- Checks for validity.
- Checks availability on CRAN, Bioconductor, and beyond.
- Searches various websites to help you discover any unintended meanings. In an interactive session, the URLs displayed by available are opened in browser tabs.
- Attempts to report whether the name has positive or negative sentiment.

`pak::pkg_name_check()` is alternative function with a similar purpose. Since the pak package is under more active development than available, it may emerge as the better option going forward.

Package Creation

Once you've come up with a name, there are two ways to create the package:

- Call `usethis::create_package()`.

- In RStudio, do *File > New Project > New Directory > R Package*. This ultimately calls `usethis::create_package()`, so really there's just one way.

This produces the smallest possible *working* package, with three components:

- An *R/* directory, which you'll learn about in Chapter 6.
- A basic *DESCRIPTION* file, which you'll learn about in Chapter 9.
- A basic *NAMESPACE* file, which you'll learn about in "The NAMESPACE File" on page 144.

It may also include an RStudio project file, *pkgname.Rproj*, that makes your package easy to use with RStudio, as described in "RStudio Projects" on page 52. Basic *.Rbuildignore* and *.gitignore* files are also left behind.

 Don't use `package.skeleton()` to create a package. Because this function comes with R, you might be tempted to use it, but it creates a package that immediately throws errors with R CMD build. It anticipates a different development process than we use here, so repairing this broken initial state just makes unnecessary work for people who use devtools (and, especially, roxygen2). Use `create_package()`.

Where Should You create_package()?

The main and only required argument to `create_package()` is the path where your new package will live:

```
create_package("path/to/package/pkgname")
```

Remember that this is where your package lives in its *source* form (see "Source Package" on page 34), not in its *installed* form (see "Installed Package" on page 40). Installed packages live in a *library*, and we discussed conventional setups for libraries in "Package Libraries" on page 43.

Where should you keep source packages? The main principle is that this location should be distinct from where installed packages live. In the absence of external considerations, a typical user should designate a directory inside their home directory for R (source) packages. We discussed this with colleagues, and the source of many tidyverse packages lives inside directories like *~/rrr/*, *~/documents/tidyverse/*, *~/r/packages/*, or *~/pkg/*. Some of us use one directory for this; others divide source packages among a few directories based on their development role (contributor versus not), GitHub organization (tidyverse versus r-lib), development stage (active versus not), and so on.

These directory conventions probably reflect that we are primarily tool-builders. An academic researcher might organize their files around individual publications, whereas a data scientist might organize around data products and reports. There is no particular technical or traditional reason for one specific approach. As long as you keep a clear distinction between source and installed packages, just pick a strategy that works within your overall system for file organization, and use it consistently.

RStudio Projects

devtools works hand-in-hand with RStudio, which we believe is the best development environment for most R users. To be clear, you can use devtools without using RStudio, and you can develop packages in RStudio without using devtools. But there is a special, two-way relationship that makes it very rewarding to use devtools and RStudio together.

 RStudio

An RStudio *Project*, with a capital "P," is a regular directory on your computer that includes some (mostly hidden) RStudio infrastructure to facilitate your work on one or more *projects*, with a lowercase "p". A project might be an R package, a data analysis, a Shiny app, a book, a blog, etc.

Benefits of RStudio Projects

From "Source Package" on page 34, you already know that a source package lives in a directory on your computer. We strongly recommend that each source package is also an RStudio Project. Here are some of the payoffs:

- Projects are very "launch-able." It's easy to fire up a fresh instance of RStudio in a Project, with the file browser and working directory set exactly the way you need, ready for work.

- Each Project is isolated; code run in one Project does not affect any other Project.

  — You can have several RStudio Projects open at once, and code executed in Project A does not have any effect on the R session and workspace of Project B.

- You get handy code navigation tools like F2 to jump to a function definition and Ctrl + . to look up functions or files by name.

- You get useful keyboard shortcuts and a clickable interface for common package development tasks, like generating documentation, running tests, or checking the entire package.

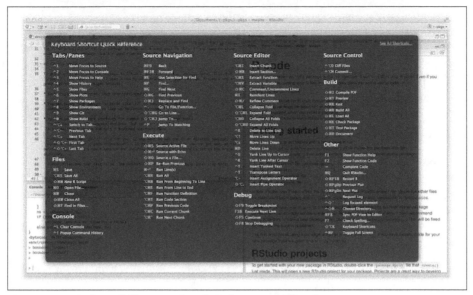

Figure 4-1. Keyboard Shortcut Quick Reference in RStudio

To see the most useful keyboard shortcuts, press Alt+Shift+K or use *Help > Keyboard Shortcuts Help*. You should see something like Figure 4-1.

RStudio

RStudio also provides the *Command Palette* (*https://oreil.ly/ Hn8hA*), which gives fast, searchable access to all of the IDE's commands. This is especially helpful when you can't remember a particular keyboard shortcut. It is invoked via Ctrl+Shift+P (Windows & Linux) or Cmd-Shift-P (macOS).

RStudio

Follow @rstudiotips (*https://twitter.com/rstudiotips*) on Twitter for a regular dose of RStudio tips and tricks.

How to Get an RStudio Project

If you follow our recommendation to create new packages with `create_package()`, each new package will also be an RStudio Project, if you're working from RStudio.

If you need to designate the directory of a preexisting source package as an RStudio Project, choose one of these options:

- In RStudio, do *File > New Project > Existing Directory*.
- Call `create_package()` with the path to the preexisting R source package.
- Call `usethis::use_rstudio()` with the active usethis project set to an existing R package. In practice, this probably means you just need to make sure your working directory is inside the preexisting package directory.

What Makes an RStudio Project?

A directory that is an RStudio Project will contain an *.Rproj* file. Typically, if the directory is named "foo" the Project file is *foo.Rproj*. And if that directory is also an R package, then the package name is usually also "foo." The path of least resistance is to make all of these names coincide and to *not* nest your package inside a subdirectory inside the Project. If you settle on a different workflow, just know it may feel like you are fighting with the tools.

An *.Rproj* file is just a text file. Here is a representative project file you might see in a Project initiated via usethis:

```
Version: 1.0

RestoreWorkspace: No
SaveWorkspace: No
AlwaysSaveHistory: Default

EnableCodeIndexing: Yes
Encoding: UTF-8

AutoAppendNewline: Yes
StripTrailingWhitespace: Yes
LineEndingConversion: Posix

BuildType: Package
PackageUseDevtools: Yes
PackageInstallArgs: --no-multiarch --with-keep.source
PackageRoxygenize: rd,collate,namespace
```

You don't need to modify this file by hand. Instead, use the interface available via *Tools > Project Options* (Figure 4-2) or *Project Options* in the Projects menu in the top-right corner (Figure 4-3).

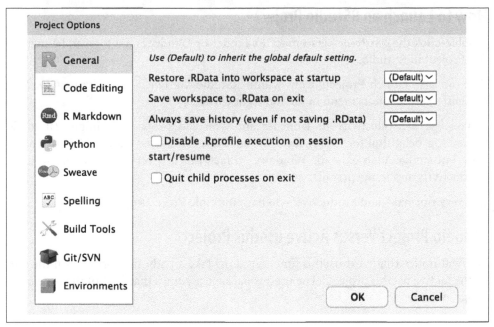

Figure 4-2. Project Options in RStudio

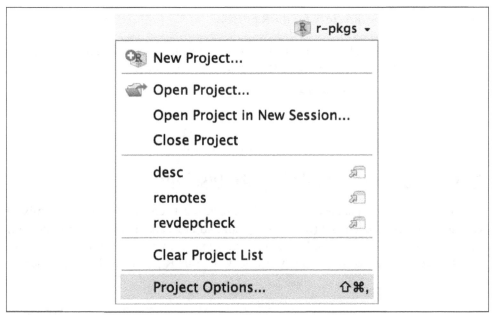

Figure 4-3. Projects Menu in RStudio

How to Launch an RStudio Project

Double-click the *foo.Rproj* file in macOS's Finder or Windows Explorer to launch the foo Project in RStudio.

You can also launch Projects from within RStudio via *File > Open Project (in New Session)* or the Projects menu in the top-right corner.

If you use a productivity or launcher app, you can probably configure it to do something delightful for *.Rproj* files. We both use Alfred for this,[2] which is macOS only, but similar tools exist for Windows. In fact, this is a very good reason to use a productivity app in the first place.

It is very normal—and productive!—to have multiple Projects open at once.

RStudio Project Versus Active usethis Project

You will notice that most usethis functions don't take a path: they operate on the files in the "active usethis project." The usethis package assumes that 95% of the time all of these coincide:

- The current RStudio Project, if using RStudio.
- The active usethis project.
- Current working directory for the R process.

If things seem funky, call `proj_sitrep()` to get a "situation report." This will identify peculiar situations and propose ways to get back to a happier state:

```
# these should usually be the same (or unset)
proj_sitrep()
#> *    working_directory: '/Users/jenny/rrr/readxl'
#> * active_usethis_proj: '/Users/jenny/rrr/readxl'
#> * active_rstudio_proj: '/Users/jenny/rrr/readxl'
```

Working Directory and Filepath Discipline

As you develop your package, you will be executing R code. This will be a mix of workflow calls (e.g., `document()` or `test()`) and ad hoc calls that help you write your functions, examples, and tests. We *strongly recommend* that you keep the top-level of your source package as the working directory of your R process. This will generally

---

2 Specifically, we configure Alfred to favor *.Rproj* files in its search results when proposing apps or files to open. To register the *.Rproj* file type with Alfred, go to *Preferences > Features > Default Results > Advanced*. Drag any *.Rproj* file onto this space and then close.

happen by default, so this is really a recommendation to avoid development workflows that require you to fiddle with the working directory.

If you're totally new to package development, you don't have much basis for supporting or resisting this proposal. But those with some experience may find this recommendation somewhat upsetting. You may be wondering how you are supposed to express paths when working in subdirectories, such as *tests/*. As it becomes relevant, we'll show you how to exploit path-building helpers, such as `testthat::test_path()`, that determine paths at execution time.

The basic idea is that by leaving the working directory alone, you are encouraged to write paths that convey intent explicitly ("read *foo.csv* from the test directory") instead of implicitly ("read *foo.csv* from current working directory, which I *think* is going to be the test directory"). A sure sign of reliance on implicit paths is incessant fiddling with your working directory, because you're using `setwd()` to manually fulfill the assumptions that are implicit in your paths.

Using explicit paths can design away a whole class of path headaches and makes day-to-day development more pleasant as well. There are two reasons why implicit paths are hard to get right:

- Recall the different forms that a package can take during the development cycle (Chapter 3). These states differ from each other in terms of which files and folders exist and their relative positions within the hierarchy. It's tricky to write relative paths that work across all package states.

- Eventually, your package will be processed with built-in tools like `R CMD build`, `R CMD check`, and `R CMD INSTALL`, by you and potentially CRAN. It's hard to keep track of what the working directory will be at every stage of these processes.

Path helpers like `testthat::test_path()`, `fs::path_package()`, and the rprojroot package (*https://rprojroot.r-lib.org*) are extremely useful for building resilient paths that hold up across the whole range of situations that come up during development and usage. Another way to eliminate brittle paths is to be rigorous in your use of proper methods for storing data inside your package (Chapter 7) and to target the session temp directory when appropriate, such as for ephemeral testing artifacts (Chapter 13).

Test Drive with load_all()

The `load_all()` function is arguably the most important part of the devtools workflow:

```
# with devtools attached and
# working directory set to top-level of your source package ...

load_all()

# ... now experiment with the functions in your package
```

`load_all()` is the key step in this "lather, rinse, repeat" cycle of package development:

1. Tweak a function definition.
2. `load_all()`.
3. Try out the change by running a small example or some tests.

When you're new to package development or to devtools, it's easy to overlook the importance of `load_all()` and fall into some awkward habits from a data analysis workflow.

Benefits of load_all()

When you first start to use a development environment, like RStudio or VS Code, the biggest win is the ability to send lines of code from an *.R* script for execution in R console. The fluidity of this is what makes it tolerable to follow the best practice of regarding your source code as real[3] (as opposed to objects in the workspace) and saving *.R* files (as opposed to saving and reloading *.Rdata*).

`load_all()` has the same significance for package development and, ironically, requires that you *not* test drive package code in the same way as script code. `load_all()` *simulates* the full-blown process for seeing the effect of a source code change, which is clunky enough[4] that you won't want to do it very often. Figure 4-4 reinforces that the `library()` function can only load a package that has been installed, whereas `load_all()` gives a high-fidelity simulation of this, based on the current package source.

---

3 Quoting the usage philosophy favored by Emacs Speaks Statistics (*https://oreil.ly/Tk3Nn*) (ESS).

4 The command-line approach is to quit R, go to the shell, do R CMD build foo in the package's parent directory, then R CMD INSTALL foo_x.y.x.tar.gz, restart R, and call library(foo).

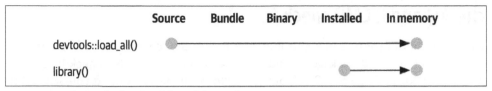

	Source	Bundle	Binary	Installed	In memory
devtools::load_all()					
library()					

Figure 4-4. devtools::load_all() vs. library()

The main benefits of load_all() include:

- You can iterate quickly, which encourages exploration and incremental progress.
  - This iterative speedup is especially noticeable for packages with compiled code.
- You get to develop interactively under a namespace regime that accurately mimics how things are when someone uses your installed package, with the following additional advantages:
  - You can call your own internal functions directly, without using ::: and without being tempted to temporarily define your functions in the global workspace.
  - You can also call functions from other packages that you've imported into your *NAMESPACE*, without being tempted to attach these dependencies via library().

load_all() removes friction from the development workflow and eliminates the temptation to use workarounds that often lead to mistakes around namespace and dependency management.

Other Ways to Call load_all()

devtools::load_all() is a thin wrapper around pkgload::load_all() that adds a bit of user-friendliness. It is unlikely you will use load_all() programmatically or inside another package, but if you do, you should probably use pkgload::load_all() directly.

RStudio

When working in a Project that is a package, RStudio offers several ways to call load_all():

- Use the keyboard shortcuts: Cmd-Shift-L (macOS) or Ctrl+Shift+L (Windows, Linux)
- Use the Build pane's More menu
- Use the Build > Load All menu option

check() and R CMD check

Base R provides various command-line tools and R CMD check is the official method for checking that an R package is valid. It is essential to pass R CMD check if you plan to submit your package to CRAN, but we *highly recommend* holding yourself to this standard even if you don't intend to release your package on CRAN. R CMD check detects many common problems that you'd otherwise discover the hard way.

Our recommended way to run R CMD check is in the R console via devtools:

```
devtools::check()
```

We recommend this because it allows you to run R CMD check from within R, which dramatically reduces friction and increases the likelihood that you will check() early and often! This emphasis on fluidity and fast feedback is exactly the same motivation as given for load_all(). In the case of check(), it really is executing R CMD check for you. It's not just a high fidelity simulation, which is the case for load_all().

 RStudio

RStudio exposes check() in the *Build* menu, in the *Build* pane via *Check*, and in keyboard shortcuts Ctrl+Shift+E (Windows & Linux) or Cmd-Shift-E (macOS).

A rookie mistake that we see often in new package developers is to do too much work on their package before running R CMD check. Then, when they do finally run it, it's typical to discover many problems, which can be very demoralizing. It's counterintuitive, but the key to minimizing this pain is to run R CMD check more often: the sooner you find a problem, the easier it is to fix.[5] We model this behavior very intentionally in Chapter 1.

The upper limit of this approach is to run R CMD check every time you make a change. We don't run check() manually quite that often, but when we're actively working on a package, it's typical to check() multiple times per day. Don't tinker with your package for days, weeks, or months, waiting for some special milestone to finally run R CMD check. If you use GitHub (see "Git and GitHub" on page 296), we'll show you how to set things up so that R CMD check runs automatically every time you push (see "GitHub Actions" on page 298).

---

5 A great blog post advocating for "if it hurts, do it more often" is FrequencyReducesDifficulty (*https://oreil.ly/RMDBG*) by Martin Fowler.

Workflow

Here's what happens inside `devtools::check()`:

- Ensures that the documentation is up-to-date by running `devtools::document()`.

- Bundles the package before checking it (see "Bundled Package" on page 35). This is the best practice for checking packages because it makes sure the check starts with a clean slate: because a package bundle doesn't contain any of the temporary files that can accumulate in your source package, e.g., artifacts like *.so* and *.o* files that accompany compiled code, you can avoid the spurious warnings such files will generate.

- Sets the `NOT_CRAN` environment variable to `"true"`. This allows you to selectively skip tests on CRAN. See `?testthat::skip_on_cran` and "Skip a Test" on page 228 for details.

The workflow for checking a package is simple but tedious:

1. Run `devtools::check()`, or press Ctrl/Cmd-Shift-E.
2. Fix the first problem.
3. Repeat until there are no more problems.

`R CMD check` returns three types of messages:

ERRORs
: Severe problems that you should fix regardless of whether you're submitting to CRAN.

WARNINGs
: Likely problems that you must fix if you're planning to submit to CRAN (and a good idea to look into even if you're not).

NOTEs
: Mild problems or, in a few cases, just an observation. If you are submitting to CRAN, you should strive to eliminate all NOTEs, even if they are false positives. If you have no NOTEs, human intervention is not required, and the package submission process will be easier. If it's not possible to eliminate a NOTE, you'll need describe why it's OK in your submission comments, as described in "The Submission Process" on page 336. If you're not submitting to CRAN, carefully read each NOTE. If it's easy to eliminate the NOTEs, it's worth it, so that you can continue to strive for a totally clean result. But if eliminating a NOTE will have a net negative impact on your package, it is reasonable to just tolerate it. Make sure that doesn't lead to you ignoring other issues that really should be addressed.

R CMD check consists of dozens of individual checks, and it would be overwhelming to enumerate them here. See our online-only guide to R CMD check (*https://r-pkgs.org/R-CMD-check.html*) for details.

Background on R CMD check

As you accumulate package development experience, you might want to access R CMD check directly at some point. Remember that R CMD check is something you must run in the terminal, not in the R console. You can see its documentation like so:

```
R CMD check --help
```

R CMD check can be run on a directory that holds an R package in source form (see "Source Package" on page 34) or, preferably, on a package bundle (see "Bundled Package" on page 35):

```
R CMD build somepackage
R CMD check somepackage_0.0.0.9000.tar.gz
```

To learn more, see the "Checking packages" section of *Writing R Extensions* (*https://oreil.ly/l5SmF*).

## CHAPTER 5

The Package Within

This part of the book ends the same way it started, with the development of a small toy package. Chapter 1 established the basic mechanics, workflow, and tooling of package development, but it said practically nothing about the R code inside the package. This chapter focuses primarily on the package's R code and how it differs from R code in a script.

Starting with a data analysis script, you learn how to find the package that lurks within. You'll isolate and then extract reusable data and logic from the script, put this into an R package, and then use that package in a much simplified script. We've included a few rookie mistakes along the way, in order to highlight special considerations for the R code inside a package.

Note that the section headers incorporate the NATO phonetic alphabet (alfa, bravo, etc.) and have no specific meaning. They are just a convenient way to mark our progress toward a working package. It is fine to follow along just by reading, and this chapter is completely self-contained (i.e., it's not a prerequisite for material later in the book). But if you wish to see the state of specific files along the way, they can be found in the source files for the book (*https://oreil.ly/_gE6e*).

Alfa: A Script That Works

Let's consider *data-cleaning.R*, a fictional data analysis script for a group that collects reports from people who went for a swim:

> Where did you swim and how hot was it outside?

Their data usually comes as a CSV file, such as *swim.csv*:

```
name,where,temp
Adam,beach,95
Bess,coast,91
Cora,seashore,28
Dale,beach,85
Evan,seaside,31
```

data-cleaning.R begins by reading *swim.csv* into a data frame:

```
infile <- "swim.csv"
(dat <- read.csv(infile))

#>    name    where temp
#> 1 Adam    beach   95
#> 2 Bess    coast   91
#> 3 Cora seashore   28
#> 4 Dale    beach   85
#> 5 Evan  seaside   31
```

They then classify each observation as using American ("US") or British ("UK") English, based on the word chosen to describe the sandy place where the ocean and land meet. The `where` column is used to build the new `english` column:

```
dat$english[dat$where == "beach"] <- "US"
dat$english[dat$where == "coast"] <- "US"
dat$english[dat$where == "seashore"] <- "UK"
dat$english[dat$where == "seaside"] <- "UK"
```

Sadly, the temperatures are often reported in a mix of Fahrenheit and Celsius. In the absence of better information, they guess that Americans report temperatures in Fahrenheit and therefore those observations are converted to Celsius:

```
dat$temp[dat$english == "US"] <- (dat$temp[dat$english == "US"] - 32) * 5/9
dat
#>    name    where temp english
#> 1 Adam    beach 35.0      US
#> 2 Bess    coast 32.8      US
#> 3 Cora seashore 28.0      UK
#> 4 Dale    beach 29.4      US
#> 5 Evan  seaside 31.0      UK
```

Finally, this cleaned (cleaner?) data is written back out to a CSV file. They like to capture a timestamp in the filename when they do this[1]:

```
now <- Sys.time()
timestamp <- format(now, "%Y-%B-%d_%H-%M-%S")
(outfile <- paste0(timestamp, "_", sub("(.*)([.]csv$)", "\\1_clean\\2", infile)))
#> [1] "2023-March-31_11-20-41_swim_clean.csv"
write.csv(dat, file = outfile, quote = FALSE, row.names = FALSE)
```

Here is *data-cleaning.R* in its entirety:

```
infile <- "swim.csv"
(dat <- read.csv(infile))

dat$english[dat$where == "beach"] <- "US"
dat$english[dat$where == "coast"] <- "US"
dat$english[dat$where == "seashore"] <- "UK"
dat$english[dat$where == "seaside"] <- "UK"

dat$temp[dat$english == "US"] <- (dat$temp[dat$english == "US"] - 32) * 5/9
dat

now <- Sys.time()
timestamp <- format(now, "%Y-%B-%d_%H-%M-%S")
(outfile <- paste0(timestamp, "_", sub("(.*)([.]csv$)", "\\1_clean\\2", infile)))
write.csv(dat, file = outfile, quote = FALSE, row.names = FALSE)
```

Even if your typical analytical tasks are quite different, hopefully you see a few familiar patterns here. It's easy to imagine that this group does very similar preprocessing of many similar data files over time. Their analyses can be more efficient and consistent if they make these standard data maneuvers available to themselves as functions in a package, instead of inlining the same data and logic into dozens or hundreds of data ingest scripts.

Bravo: A Better Script That Works

The package that lurks within the original script is actually pretty hard to see! It's obscured by a few suboptimal coding practices, such as the use of repetitive copy/paste-style code and the mixing of code and data. Therefore a good first step is to refactor this code, isolating as much data and logic as possible in proper objects and functions, respectively.

This is also a good time to introduce the use of some add-on packages, for several reasons. First, we would actually use the tidyverse for this sort of data wrangling.

---

1 Sys.time() returns an object of class POSIXct; therefore, when we call format() on it, we are actually using format.POSIXct(). Read the help for ?format.POSIXct (*https://rdrr.io/r/base/strptime.html*) if you're not familiar with such format strings.

Second, many people use add-on packages in their scripts, so it is good to see how add-on packages are handled inside a package.

Here's the new and improved version of the script:

```
library(tidyverse)

infile <- "swim.csv"
dat <- read_csv(infile, col_types = cols(name = "c", where = "c", temp = "d"))

lookup_table <- tribble(
     ~where,    ~english,
     "beach",    "US",
     "coast",    "US",
   "seashore",   "UK",
    "seaside",   "UK"
)

dat <- dat %>%
  left_join(lookup_table)

f_to_c <- function(x) (x - 32) * 5/9

dat <- dat %>%
  mutate(temp = if_else(english == "US", f_to_c(temp), temp))
dat

now <- Sys.time()
timestamp <- function(time) format(time, "%Y-%B-%d_%H-%M-%S")
outfile_path <- function(infile) {
  paste0(timestamp(now), "_", sub("(.*)([.]csv$)", "\\1_clean\\2", infile))
}
write_csv(dat, outfile_path(infile))
```

The key changes to note are:

- We are using functions from tidyverse packages (specifically from readr and dplyr), and we make them available with library(tidyverse).

- The map between different "beach" words and whether they are considered to be US or UK English is now isolated in a lookup table, which lets us create the english column in one go with a left_join(). This lookup table makes the mapping easier to comprehend and would be much easier to extend in the future with new "beach" words.

- f_to_c(), timestamp(), and outfile_path() are new helper functions that hold the logic for converting temperatures and forming the timestamped output file name.

It's getting easier to recognize the reusable bits of this script, i.e., the bits that have nothing to do with a specific input file, like *swim.csv*. This sort of refactoring often happens naturally on the way to creating your own package, but if it does not, it's a good idea to do this intentionally.

Charlie: A Separate File for Helper Functions

A typical next step is to move reusable data and logic out of the analysis script and into one or more separate files. This is a conventional opening move, if you want to use these same helper files in multiple analyses.

Here is the content of *beach-lookup-table.csv*:

```
where,english
beach,US
coast,US
seashore,UK
seaside,UK
```

Here is the content of *cleaning-helpers.R*:

```
library(tidyverse)

localize_beach <- function(dat) {
  lookup_table <- read_csv(
    "beach-lookup-table.csv",
    col_types = cols(where = "c", english = "c")
  )
  left_join(dat, lookup_table)
}

f_to_c <- function(x) (x - 32) * 5/9

celsify_temp <- function(dat) {
  mutate(dat, temp = if_else(english == "US", f_to_c(temp), temp))
}

now <- Sys.time()
timestamp <- function(time) format(time, "%Y-%B-%d_%H-%M-%S")
outfile_path <- function(infile) {
  paste0(timestamp(now), "_", sub("(.*)([.]csv$)", "\\1_clean\\2", infile))
}
```

We've added some high-level helper functions, `localize_beach()` and `cel sify_temp()`, to the preexisting helpers (`f_to_c()`, `timestamp()`, and `out file_path()`).

Here is the next version of the data cleaning script, now that we've pulled out the helper functions (and lookup table):

```
library(tidyverse)
source("cleaning-helpers.R")

infile <- "swim.csv"
dat <- read_csv(infile, col_types = cols(name = "c", where = "c", temp = "d"))

(dat <- dat %>%
    localize_beach() %>%
    celsify_temp())

write_csv(dat, outfile_path(infile))
```

Notice that the script is getting shorter and, hopefully, easier to read and modify, because repetitive and fussy clutter has been moved out of sight. Whether the code is actually easier to work with is subjective and depends on how natural the "interface" feels for the people who actually preprocess swimming data. These sorts of design decisions are the subject of a separate project (*https://design.tidyverse.org*).

Let's assume the group agrees that our design decisions are promising, i.e., we seem to be making things better, not worse. Sure, the existing code is not perfect, but this is a typical developmental stage when you're trying to figure out what the helper functions should be and how they should work.

Delta: A Failed Attempt at Making a Package

While this first attempt to create a package will end in failure, it's still helpful to go through some common missteps, to illuminate what happens behind the scenes.

Here are the simplest steps that you might take, in an attempt to convert *cleaning-helpers.R* into a proper package:

- Use usethis::create_package("path/to/delta") to scaffold a new R package, with the name "delta."

  — This is a good first step!

- Copy *cleaning-helpers.R* into the new package, specifically, to *R/cleaning-helpers.R*.

  — This is morally correct, but mechanically wrong in several ways, as we will soon see.

- Copy *beach-lookup-table.csv* into the new package. But where? Let's try the top-level of the source package.
  - This is not going to end well. Shipping data files in a package is a special topic, which is covered in Chapter 7.
- Install this package, perhaps using `devtools::install()` or via Ctrl+Shift+B (Windows and Linux) or Cmd-Shift-B (macOS) in RStudio.
  - Despite all of the problems previously identified, this actually works! Which is interesting, because we can (try to) use it and see what happens.

Here is the next version of the data cleaning script that you hope will run after successfully installing this package (which we're calling "delta"):

```
library(tidyverse)
library(delta)

infile <- "swim.csv"
dat <- read_csv(infile, col_types = cols(name = "c", where = "c", temp = "d"))

dat <- dat %>%
  localize_beach() %>%
  celsify_temp()

write_csv(dat, outfile_path(infile))
```

The only change from our previous script is that

```
source("cleaning-helpers.R")
```

has been replaced by

```
library(delta)
```

Here's what actually happens if you install the delta package and try to run the data cleaning script:

```
library(tidyverse)
library(delta)

infile <- "swim.csv"
dat <- read_csv(infile, col_types = cols(name = "c", where = "c", temp = "d"))

dat <- dat %>%
  localize_beach() %>%
  celsify_temp()
#> Error in localize_beach(.) : could not find function "localize_beach"

write_csv(dat, outfile_path(infile))
#> Error in outfile_path(infile) : could not find function "outfile_path"
```

None of the helper functions are actually available for use, even though you call library(delta)! In contrast to source()ing a file of helper functions, attaching a package does not dump its functions into the global workspace. By default, functions in a package are only for internal use. You need to export localize_beach(), celsify_temp(), and outfile_path() so your users can call them. In the devtools workflow, we achieve this by putting @export in the special roxygen comment above each function (namespace management is covered in "NAMESPACE Workflow" on page 156), like so:

```
#' @export
celsify_temp <- function(dat) {
    mutate(dat, temp = if_else(english == "US", f_to_c(temp), temp))
}
```

After you add the @export tag to localize_beach(), celsify_temp(), and outfile_path(), you run devtools::document() to (re)generate the *NAMESPACE* file, and reinstall the delta package. Now when you re-execute the data cleaning script, it works!

Correction: it *sort of* works *sometimes*. Specifically, it works if and only if the working directory is set to the top-level of the source package. From any other working directory, you still get an error:

```
dat <- dat %>%
  localize_beach() %>%
  celsify_temp()
#> Error: 'beach-lookup-table.csv' does not exist in current working directory
#> ('/Users/jenny/tmp').
```

The lookup table consulted inside localize_beach() cannot be found. One does not simply dump CSV files into the source of an R package and expect things to "just work." We will fix this in our next iteration of the package (Chapter 7 has full coverage of how to include data in a package).

Before we abandon this initial experiment, let's also marvel at the fact that you were able to install, attach, and, to a certain extent, use a fundamentally broken package. devtools::load_all() works fine, too! This is a sobering reminder that you should be running R CMD check, probably via devtools::check(), very often during development. This will quickly alert you to many problems that simple installation and usage does not reveal.

Indeed, check() fails for this package and you see this:

```
* installing _source_ package 'delta' ...
** using staged installation
** R
** byte-compile and prepare package for lazy loading
Error in library(tidyverse) : there is no package called 'tidyverse'
Error: unable to load R code in package 'delta'
Execution halted
ERROR: lazy loading failed for package 'delta'
* removing '/Users/jenny/rrr/delta.Rcheck/delta'
```

What do you mean "there is no package called 'tidyverse'"?!? We're using it, with no problems, in our main script! Also, we've already installed and used this package, why can't R CMD check find it?

This error is what happens when the strictness of R CMD check meets the very first line of *R/cleaning-helpers.R*:

```
library(tidyverse)
```

This is *not* how you declare that your package depends on another package (the tidyverse, in this case). This is *also not* how you make functions in another package available for use in yours. Dependencies must be declared in DESCRIPTION (and that's not all). Since we declared no dependencies, R CMD check takes us at our word and tries to install our package with only the base packages available, which means this library(tidyverse) call fails. A "regular" installation succeeds, simply because the tidyverse is available in your regular library, which hides this particular mistake.

To review, copying *cleaning-helpers.R* to *R/cleaning-helpers.R*, without further modification, was problematic in (at least) the following ways:

- Does not account for exported versus nonexported functions.
- The CSV file holding our lookup table cannot be found in the installed package.
- Does not properly declare our dependency on other add-on packages.

Echo: A Working Package

We're ready to make the most minimal version of this package that actually works.

Here is the new version of *R/cleaning-helpers.R:*[2]

```
lookup_table <- dplyr::tribble(
    ~where, ~english,
    "beach",    "US",
```

---

2 Putting everything in one file, with this name, is not ideal, but it is technically allowed. We discuss organizing and naming the files below *R/* in "Organize Functions Into Files" on page 83.

```
      "coast",       "US",
    "seashore",      "UK",
      "seaside",     "UK"
)

#' @export
localize_beach <- function(dat) {
  dplyr::left_join(dat, lookup_table)
}

f_to_c <- function(x) (x - 32) * 5/9

#' @export
celsify_temp <- function(dat) {
  dplyr::mutate(dat, temp = dplyr::if_else(english == "US", f_to_c(temp), temp))
}

now <- Sys.time()
timestamp <- function(time) format(time, "%Y-%B-%d_%H-%M-%S")

#' @export
outfile_path <- function(infile) {
  paste0(timestamp(now), "_", sub("(.*)([.]csv$)", "\\1_clean\\2", infile))
}
```

We've gone back to defining the `lookup_table` with R code, since the initial attempt to read it from CSV created some sort of filepath snafu. This is OK for small, internal, static data, but remember to see Chapter 7 for more general techniques for storing data in a package.

All of the calls to tidyverse functions have now been qualified with the name of the specific package that actually provides the function, e.g., `dplyr::mutate()`. There are other ways to access functions in another package, explained in "Package Is Listed in Imports" on page 157, but this is our recommended default. It is also our strong recommendation that no one depend on the tidyverse metapackage in a package.[3] Instead, it is better to identify the specific package(s) you actually use. In this case, the package only uses dplyr.

The `library(tidyverse)` call is gone and instead we declare the use of dplyr in the `Imports` field of `DESCRIPTION`:

```
Package: echo
(... other lines omitted ...)
Imports:
    dplyr
```

---

3 The blog post "The tidyverse is for EDA, not packages" (*https://oreil.ly/xZIGP*) elaborates on this.

This, together with the use of namespace-qualified calls, like dplyr::left_join(), constitutes a valid way to use another package within yours. The metadata conveyed via *DESCRIPTION* is covered in Chapter 9.

All of the user-facing functions have an @export tag in their roxygen comment, which means that devtools::document() adds them correctly to the *NAMESPACE* file. Note that f_to_c() is currently used only internally, inside celsify_temp(), so it is not exported (likewise for timestamp()).

This version of the package can be installed, used, *and* it technically passes R CMD check, though with 1 warning and 1 note:

```
* checking for missing documentation entries ... WARNING
Undocumented code objects:
  'celsify_temp' 'localize_beach' 'outfile_path'
All user-level objects in a package should have documentation entries.
See chapter 'Writing R documentation files' in the 'Writing R
Extensions' manual.

* checking R code for possible problems ... NOTE
celsify_temp: no visible binding for global variable 'english'
celsify_temp: no visible binding for global variable 'temp'
Undefined global functions or variables:
  english temp
```

The "no visible binding" note is a peculiarity of using dplyr and unquoted variable names inside a package, where the use of bare variable names (english and temp) looks suspicious. You can add either of these lines to any file below *R/* to eliminate this note (such as the package-level documentation file described in "Help Topic for the Package" on page 256):

```
# option 1 (then you should also put utils in Imports)
utils::globalVariables(c("english", "temp"))

# option 2
english <- temp <- NULL
```

We're seeing that it can be tricky to program around a package like dplyr, which makes heavy use of nonstandard evaluation. Behind the scenes, that is the technique that allows dplyr's end users to use bare (not quoted) variable names. Packages like dplyr prioritize the experience of the typical end user, at the expense of making them trickier to depend on. The two options given for suppressing the "no visible binding" note represent entry-level solutions. For a more sophisticated treatment of these issues, see vignette("in-packages", package = "dplyr") and vignette("programming", package = "dplyr").

The warning about missing documentation is because the exported functions have not been properly documented. This is a valid concern and something you absolutely should address in a real package. You've already seen how to create help files with roxygen comments in "document()" on page 12, and we cover this topic thoroughly in Chapter 16.

Foxtrot: Build Time Versus Run Time

The echo package works, which is great, but group members notice something odd about the timestamps:

```
Sys.time()
#> [1] "2023-03-26 22:48:48 PDT"

outfile_path("INFILE.csv")
#> [1] "2020-September-03_11-06-33_INFILE_clean.csv"
```

The datetime in the timestamped filename doesn't reflect the time reported by the system. In fact, the users claim that the timestamp never seems to change at all! Why is this?

Recall how we form the filepath for output files:

```
now <- Sys.time()
timestamp <- function(time) format(time, "%Y-%B-%d_%H-%M-%S")
outfile_path <- function(infile) {
  paste0(timestamp(now), "_", sub("(.*)([.]csv$)", "\\1_clean\\2", infile))
}
```

The fact that we capture `now <- Sys.time()` outside of the definition of `out file_path()` has probably been vexing some readers for a while. `now` reflects the instant in time when we execute `now <- Sys.time()`. In the initial approach, `now` was assigned when we `source()`d *cleaning-helpers.R*. That's not ideal, but it was probably a pretty harmless mistake, because the helper file would be `source()`d shortly before we wrote the output file.

But this approach is quite devastating in the context of a package. `now <- Sys.time()` is executed when the package is built.[4] And never again. It is very easy to assume your package code is reevaluated when the package is attached or used. But it is not. Yes, absolutely, the code *inside your functions* is run whenever they are called. But your functions—and any other objects created in top-level code below *R/*—are defined exactly once, at build time.

---

4 Here we're referring to when the package code is compiled, which could be either when the binary is made (for macOS or Windows; see "Binary Package" on page 38) or when the package is installed from source "Installed Package" on page 40.

By defining now with top-level code below *R/*, we've doomed our package to time-stamp all of its output files with the same (wrong) time. The fix is to make sure the `Sys.time()` call happens at run time.

Let's look again at parts of *R/cleaning-helpers.R*:

```
lookup_table <- dplyr::tribble(
    ~where,    ~english,
    "beach",      "US",
    "coast",      "US",
  "seashore",     "UK",
   "seaside",     "UK"
)

now <- Sys.time()
timestamp <- function(time) format(time, "%Y-%B-%d_%H-%M-%S")
outfile_path <- function(infile) {
  paste0(timestamp(now), "_", sub("(.*)([.]csv$)", "\\1_clean\\2", infile))
}
```

There are four top-level `<-` assignments in this excerpt. The top-level definitions of the data frame `lookup_table` and the functions `timestamp()` and `outfile_path()` are correct. It is appropriate that these be defined exactly once, at build time. The top-level definition of now, which is then used inside `outfile_path()`, is regrettable.

Here are better versions of `outfile_path()`:

```
# always timestamp as "now"
outfile_path <- function(infile) {
  ts <- timestamp(Sys.time())
  paste0(ts, "_", sub("(.*)([.]csv$)", "\\1_clean\\2", infile))
}

# allow user to provide a time, but default to "now"
outfile_path <- function(infile, time = Sys.time()) {
  ts <- timestamp(time)
  paste0(ts, "_", sub("(.*)([.]csv$)", "\\1_clean\\2", infile))
}
```

This illustrates that you need to have a different mindset when defining objects inside a package. The vast majority of those objects should be functions, and these functions should generally only use data they create or that is passed via an argument. There are some types of sloppiness that are fairly harmless when a function is defined immediately before its use, but that can be more costly for functions distributed as a package.

Golf: Side Effects

The timestamps now reflect the current time, but the group raises a new concern. As it stands, the timestamps reflect who has done the data cleaning and which part of the world they're in. The heart of the timestamp strategy is this format string:[5]

```
format(Sys.time(), "%Y-%B-%d_%H-%M-%S")
#> [1] "2023-March-31_11-20-42"
```

This formats `Sys.time()` in such a way that it includes the month *name* (not number) and the local time.[6]

Table 5-1 shows what happens when such a timestamp is produced by several hypothetical colleagues cleaning some data at exactly the same instant in time.

Table 5-1. Timestamp varies by locale and time zone

Location	Timestamp	LC_TIME	tz
Rome, Italy	2020-Settembre-05_00-30-00	it_IT.UTF-8	Europe/Rome
Warsaw, Poland	2020-września-05_00-30-00	pl_PL.UTF-8	Europe/Warsaw
Sao Paulo, Brazil	2020-Setembro-04_19-30-00	pt_BR.UTF-8	America/Sao_Paulo
Greenwich, England	2020-September-04_23-30-00	en_GB.UTF-8	Europe/London
"Computer World!"	2020-September-04_22-30-00	C	UTC

Note that the month names vary, as does the time, and even the date! The safest choice is to form timestamps with respect to a fixed locale and time zone (presumably the nongeographic choices represented by "Computer World!").

You do some research and learn that you can force a certain locale via `Sys.set locale()` and force a certain time zone by setting the TZ environment variable. Specifically, we set the LC_TIME component of the locale to "C" and the time zone to "UTC" (Coordinated Universal Time). Here's your first attempt to improve `timestamp()`:

```
timestamp <- function(time = Sys.time()) {
  Sys.setlocale("LC_TIME", "C")
  Sys.setenv(TZ = "UTC")
  format(time, "%Y-%B-%d_%H-%M-%S")
}
```

---

5 `Sys.time()` returns an object of class `POSIXct`; therefore, when we call `format()` on it, we are actually using `format.POSIXct()`. Read the help for `?format.POSIXct` (*https://rdrr.io/r/base/strptime.html*) if you're not familiar with such format strings.

6 It would clearly be better to format according to ISO 8601, which encodes the month by number, but please humor us for the sake of making this example more obvious.

But your Brazilian colleague notices that datetimes print differently, before and after she uses outfile_path() from your package.

Before:

```
format(Sys.time(), "%Y-%B-%d_%H-%M-%S")

#> [1] "2023-Março-31_15-20-43"
```

After:

```
outfile_path("INFILE.csv")
#> [1] "2023-March-31_18-20-42_INFILE_clean.csv"

format(Sys.time(), "%Y-%B-%d_%H-%M-%S")
#> [1] "2023-March-31_18-20-43"
```

Notice that her month name switched from Portuguese to English and the time is clearly being reported in a different time zone. The calls to Sys.setlocale() and Sys.setenv() inside timestamp() have made persistent (and very surprising) changes to her R session. This sort of side effect is very undesirable and is extremely difficult to track down and debug, especially in more complicated settings.

Here are better versions of timestamp():

```
# use withr::local_*() functions to keep the changes local to timestamp()
timestamp <- function(time = Sys.time()) {
  withr::local_locale(c("LC_TIME" = "C"))
  withr::local_timezone("UTC")
  format(time, "%Y-%B-%d_%H-%M-%S")
}

# use the tz argument to format.POSIXct()
timestamp <- function(time = Sys.time()) {
  withr::local_locale(c("LC_TIME" = "C"))
  format(time, "%Y-%B-%d_%H-%M-%S", tz = "UTC")
}

# put the format() call inside withr::with_*()
timestamp <- function(time = Sys.time()) {
  withr::with_locale(
    c("LC_TIME" = "C"),
    format(time, "%Y-%B-%d_%H-%M-%S", tz = "UTC")
  )
}
```

These show various methods to limit the scope of our changes to LC_TIME and the time zone. A good rule of thumb is to make the scope of such changes as narrow as is possible and practical. The tz argument of format() is the most surgical way to deal with the time zone, but nothing similar exists for LC_TIME. We make the temporary locale modification using the withr package, which provides a very flexible toolkit for temporary state changes. This (and base::on.exit()) are discussed further in

"Respect the R Landscape" on page 91. Note that if you use withr as we do here, you would need to list it in DESCRIPTION in Imports (see Chapter 11, "Dependency Thoughts Specific to the tidyverse" on page 140).

This underscores a point from the previous section: you need to adopt a different mindset when defining functions inside a package. Try to avoid making any changes to the user's overall state. If such changes are unavoidable, make sure to reverse them (if possible) or to document them explicitly (if related to the function's primary purpose).

Concluding Thoughts

Finally, after several iterations, we have successfully extracted the repetitive data cleaning code for the swimming survey into an R package. This example concludes the first part of book and marks the transition into more detailed reference material on specific package components. Before we move on, let's review the lessons learned in this chapter.

Script Versus Package

When you first hear that expert R users often put their code into packages, you might wonder exactly what that means. Specifically, what happens to your existing R scripts, R Markdown reports, and Shiny apps? Does all of that code somehow get put into a package? The answer is "no," in most contexts.

Typically, you identify certain recurring operations that occur across multiple projects and this is what you extract into an R package. You will still have R scripts, R Markdown reports, and Shiny apps, but by moving specific pieces of code into a formal package, your data products tend to become more concise and easier to maintain.

Finding the Package Within

Although the example in this chapter is rather simple, it still captures the typical process of developing an R package for personal or organizational use. You typically start with a collection of idiosyncratic and related R scripts, scattered across different projects. Over time, you begin to notice that certain needs come up over and over again.

Each time you revisit a similar analysis, you might try to elevate your game a bit, compared to the previous iteration. You refactor copy/paste-style code using more robust patterns and start to encapsulate key "moves" in helper functions, which might eventually migrate into their own file. Once you reach this stage, you're in a great position to take the next step and create a package.

Package Code Is Different

Writing package code is a bit different from writing R scripts, and it's natural to feel some discomfort when making this adjustment. Here are the most common gotchas that trip many of us up at first:

- Package code requires new ways of working with functions in other packages. The *DESCRIPTION* file is the principal way to declare dependencies; we don't do this via `library(somepackage)`.

- If you want data or files to be persistently available, there are package-specific methods of storage and retrieval. You can't just put files in the package and hope for the best.

- It's necessary to be explicit about which functions are user-facing and which are internal helpers. By default, functions are not exported for use by others.

- A new level of discipline is required to ensure that code runs at the intended time (build time versus run time) and that there are no unintended side effects.

Package Components

## CHAPTER 6

R Code

The first principle of making a package is that all R code goes in the *R/* directory. In this chapter, you'll learn about organizing your functions into files, maintaining a consistent style, and recognizing the stricter requirements for functions in a package (versus in a script). We'll also remind you of the fundamental workflows for test-driving and formally checking an in-development package: load_all(), test(), and check().

Organize Functions Into Files

The only hard rule is that your package must store its function definitions in R scripts, i.e., files with extension *.R*, that live in the *R/* directory.[1] However, a few more conventions can make the source code of your package easier to navigate and relieve you of re-answering "How should I name this?" each time you create a new file. The Tidyverse Style Guide offers some general advice about filenames (*https://style.tidyverse.org/files.html*) and also advice that specifically applies to files in a package (*https://style.tidyverse.org/package-files.html*). We expand on this here.

The filename should be meaningful and convey which functions are defined within. While you're free to arrange functions into files as you wish, the two extremes are bad: don't put all functions into one file and don't put each function into its own separate file. This advice should inform your general policy, but there are exceptions to every rule. If a specific function is very large or has lots of documentation, it can make sense to give it its own file, named after the function. More often, a single *.R* file

---

1 Unfortunately you can't use subdirectories inside *R/*. The next best thing is to use a common prefix, e.g., abc-*.R, to signal that a group of files are related.

will contain multiple function definitions: such as a main function and its supporting helpers, a family of related functions, or some combination of the two.

Table 6-1 presents some examples from the actual source of the tidyr package (*http:// tidyr.tidyverse.org*) at version 1.1.2. There are some departures from the hard-and-fast rules given previously, which illustrates that there's a lot of room for judgment here.

Table 6-1. Different ways to organize functions in files

Organizing principle	Source file	Comments
One function	tidyr/R/uncount.R (*https://oreil.ly/vNeVY*)	Defines exactly one function, `uncount()`, that's not particulary large but doesn't fit naturally into any other .R file
Main function plus helpers	tidyr/R/separate.R (*https://oreil.ly/VPdlK*)	Defines the user-facing `separate()` (an S3 generic), a `data.frame` method, and private helpers
Family of functions	tidyr/R/rectangle.R (*https://oreil.ly/M8h5n*)	Defines a family of functions for "rectangling" nested lists (`hoist()` and the `unnest()` functions), all documented together in a big help topic, plus private helpers

 Another file you often see in the wild is *R/utils.R*. This is a common place to define small utilities that are used inside multiple package functions. Since they serve as helpers to multiple functions, placing them in *R/utils.R* makes them easier to rediscover when you return to your package after a long break.

Bob Rudis assembled a collection of such files and did some analysis in the post Dissecting R Package "Utility Belts" (*https://oreil.ly/ tuMhO*).

If it's very hard to predict which file a function lives in, that suggests it's time to separate your functions into more files or reconsider how you are naming your functions and/or files.

RStudio

The organization of functions within files is less important in RStudio, which offers two ways to jump to the definition of a function:

- Press Ctrl + . (the period) to bring up the *Go to File/Function* tool, as shown in Figure 6-1, then start typing the name. Keep typing to narrow the list and eventually pick a function (or file) to visit. This works for both functions and files in your project.

- With your cursor in a function name or with a function name selected, press F2. This works for functions defined in your package or in another package.

After navigating to a function with one of these methods, return to where you started by clicking the back arrow at the top left of the editor (⟵⟶) or by pressing Ctrl+F9 (Windows & Linux) or Cmd-F9 (macOS).

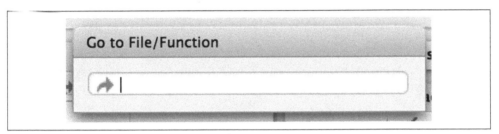

Figure 6-1. Go to File/Function in RStudio

Fast Feedback via load_all()

As you add or modify functions defined in files below *R/*, you will naturally want to try them out. We want to reiterate our strong recommendation to use devtools::load_all() to make them available for interactive exploration instead of, for example, source()ing files below *R/*. The main coverage of load_all() is in "Test Drive with load_all()" on page 58 and load_all() also shows up as one of the natural development tasks in "load_all()" on page 7. The importance of load_all() in the testthat workflow is explained in "Remove Tension Between Interactive and Automated Testing" on page 211. Compared to the alternatives, load_all() helps you to iterate more quickly and provides an excellent approximation to the namespace regime of an installed package.

Code Style

We recommend following the tidyverse style guide (*https://style.tidyverse.org*), which goes into much more detail than we can here. Its format also allows it to be a more dynamic document than this book.

Although the style guide explains the "what" and the "why," another important decision is *how* to enforce a specific code style. For this we recommend the styler package (*https://styler.r-lib.org*); its default behavior enforces the tidyverse style guide. There are many ways to apply styler to your code, depending on the context:

- `styler::style_pkg()` restyles an entire R package.
- `styler::style_dir()` restyles all files in a directory.
- `usethis::use_tidy_style()` is a wrapper that applies one of the preceding two functions depending on whether or not the current project is an R package.
- `styler::style_file()` restyles a single file.
- `styler::style_text()` restyles a character vector.

RStudio

When styler is installed, the RStudio Addins menu will offer several additional ways to style code:

- The active selection
- The active file
- The active package

If you don't use Git or another version control system, applying a function like `styler::style_pkg()` is nerve-wracking and somewhat dangerous, because you lack a way to see exactly what changed and to accept/reject such changes in a granular way.

The styler package can also be integrated with various platforms for hosting source code and doing continuous integration. For example, the tidyverse packages use a GitHub Action that restyles a package when triggered by a special comment (`/style`) on a pull request. This allows maintainers to focus on reviewing the substance of the pull request, without having to nitpick small issues of whitespace or indentation.[2][3]

Understand When Code Is Executed

Up until now, you've probably been writing *scripts*, R code saved in a file that you execute interactively, perhaps using an IDE and/or `source()`, or noninteractively via `Rscript`. There are two main differences between code in scripts and packages:

- In a script, code is run … when you run it! The awkwardness of this statement reflects that it's hard to even think about this issue with a script. However, we must, in order to appreciate that the code in a package is run *when the package is built*. This has big implications for how you write the code below *R/*: package code should only create objects, the vast majority of which will be functions.

- Functions in your package will be used in situations that you didn't imagine. This means your functions need to be thoughtful in the way that they interact with the outside world.

We expand on the first point here and the second in the next section. These topics are also illustrated concretely in "Foxtrot: Build Time Versus Run Time" on page 74.

When you `source()` a script, every line of code is executed and the results are immediately made available. Things are different with package code, because it is loaded in two steps. When the binary package is built (often, by CRAN) all the code in *R/* is executed and the results are saved. When you attach a package with `library()`, these cached results are reloaded and certain objects (mostly functions) are made available for your use. The full details on what it means for a package to be in binary form are given in "Binary Package" on page 38. We refer to the creation of the binary package as (binary) "build time" and, specifically, we mean when `R CMD INSTALL --build` is run. (You might think that this is what `R CMD build` does, but that actually makes a bundled package, a.k.a. a "source tarball.") For macOS and Windows users of CRAN packages, build time is whenever CRAN built the binary package for their OS. For those who install packages from source, build time is essentially when they (built and) installed the package.

---

2 See the Commands workflow (*https://oreil.ly/0cb-Q*) in the GitHub Actions for the R language repository (*https://github.com/r-lib/actions*).

3 The "Robot Pedantry, Human Empathy" blog post (*https://oreil.ly/wPWpp*) by Mike McQuaid does an excellent job summarizing the benefit of automating tasks like code restyling.

Consider the assignment x <- Sys.time(). If you put this in a script, x tells you when the script was source()d. But if you put that same code at the top-level in a package, x tells you when the package binary was *built*. In "Foxtrot: Build Time Versus Run Time" on page 74, we show a complete example of this in the context of forming timestamps inside a package.

The main takeaway is this:

> Any R code outside of a function is suspicious and should be carefully reviewed.

We explore a few real-world examples in the following sections that show how easy it is to get burned by this "build time versus load time" issue. Luckily, once you diagnose this problem, it is generally not difficult to fix.

Example: A Path Returned by system.file()

The shinybootstrap2 package once had this code below *R/*:

```
dataTableDependency <- list(
  htmlDependency(
    "datatables", "1.10.2",
    c(file = system.file("www/datatables", package = "shinybootstrap2")),
    script = "js/jquery.dataTables.min.js"
  ),
  htmlDependency(
    "datatables-bootstrap", "1.10.2",
    c(file = system.file("www/datatables", package = "shinybootstrap2")),
    stylesheet = c("css/dataTables.bootstrap.css", "css/dataTables.extra.css"),
    script = "js/dataTables.bootstrap.js"
  )
)
```

So dataTableDependency was a list object defined in top-level package code and its value was constructed from paths obtained via system.file(). As described in a GitHub issue (*https://oreil.ly/GvLGC*):

> This works fine when the package is built and tested on the same machine. However, if the package is built on one machine and then used on another (as is the case with CRAN binary packages), then this will fail—the dependency will point to the wrong directory on the host.

The heart of the solution is to make sure that `system.file()` is called from a function, at runtime. Indeed, this fix was made in commit 138db47 (*https://oreil.ly/ C-3GB*) and in a few other packages that had similar code and a related check was added in `htmlDependency()` itself. This particular problem would now be caught by R CMD check, due to changes that came with staged installation (*https://oreil.ly/c9s2P*) as of R 3.6.0.

Example: Available Colors

The crayon package has a function, `crayon::show_ansi_colors()`, that displays an ANSI color table on your screen, basically to show what sort of styling is possible. In an early version, the function looked something like this:

```
show_ansi_colors <- function(colors = num_colors()) {
  if (colors < 8) {
    cat("Colors are not supported")
  } else if (colors < 256) {
    cat(ansi_colors_8, sep = "")
    invisible(ansi_colors_8)
  } else {
    cat(ansi_colors_256, sep = "")
    invisible(ansi_colors_256)
  }
}

ansi_colors_8 <- # code to generate a vector covering basic terminal colors

ansi_colors_256 <- # code to generate a vector covering 256 colors
```

where `ansi_colors_8` and `ansi_colors_256` were character vectors exploring a certain set of colors, presumably styled via ANSI escapes.

The problem was those objects were formed and cached when the binary package was built. Since that often happens on a headless server, this likely happens under conditions where terminal colors might not be enabled or even available. Users of the installed package could still call `show_ansi_colors()` and `num_colors()` would detect the number of colors supported by their system (256 on most modern computers). But then an uncolored object would print to screen (the original GitHub issue is r-lib/crayon#37 (*https://oreil.ly/mgGxD*)).

The solution was to compute the display objects with a function at runtime (in commit e2b368a (*https://oreil.ly/7PSaN*)):

```
show_ansi_colors <- function(colors = num_colors()) {
  if (colors < 8) {
    cat("Colors are not supported")
  } else if (colors < 256) {
    cat(ansi_colors_8(), sep = "")
    invisible(ansi_colors_8())
  } else {
    cat(ansi_colors_256(), sep = "")
    invisible(ansi_colors_256())
  }
}

ansi_colors_8 <- function() {
  # code to generate a vector covering basic terminal colors
}

ansi_colors_256 <- function() {
  # code to generate a vector covering 256 colors
}
```

Literally, the same code is used; it is simply pushed down into the body of a function taking no arguments (similar to the shinybootstrap2 example). Each reference to, e.g., the ansi_colors_8 object is replaced by a call to the ansi_colors_8() function.

The main takeaway is that functions that assess or expose the capabilities of your package on a user's system must fully execute on your user's system. It's fairly easy to accidentally rely on results that were cached at build time, quite possibly on a different machine.

Example: Aliasing a Function

One last example shows that, even if you are careful to only define functions below *R/*, there are still some subtleties to consider. Imagine that you want the function foo() in your package to basically be an alias for the function blah() from some other package, e.g., pkgB. You might be tempted to do this:

```
foo <- pkgB::blah
```

However, this will cause `foo()` in your package to reflect the definition of `pkgB::blah()` at the version present on the machine where the binary package is built (often CRAN), at that moment in time. If a bug is discovered in `pkgB::blah()` and subsequently fixed, your package will still use the older, buggy version, until your package is rebuilt (often by CRAN) and your users upgrade, which is completely out of your control. This alternative approach protects you from this:

```
foo <- function(...) pkgB::blah(...)
```

Now, when your user calls `foo()`, they are effectively calling `pkgB::blah()`, at the version installed on *their* machine at that very moment.

A real example of this affected an older version of knitr, related to how the default "evaluate" hook was being set to `evaluate::evaluate()` (original issue is yihui/knitr#1441 (*https://oreil.ly/U4ZNy*), resolved in commit d6b53e0 (*https://oreil.ly/nX8-5*)).

Respect the R Landscape

Another big difference between a script and a package is that other people are going to use your package, and they're going to use it in situations that you never imagined. This means you need to pay attention to the R landscape, which includes not just the available functions and objects but all the global settings.

You have changed the R landscape if you've loaded a package with `library()`, or changed a global option with `options()`, or modified the working directory with `setwd()`. If the behavior of *other* functions differs before and after running your function, you've modified the landscape. "Golf: Side Effects" on page 76 has a concrete example of this involving time zones and the locale-specific printing of datetimes. Changing the landscape is bad because it makes code much harder to understand.

There are some functions that modify global settings that you should never use because there are better alternatives:

- Don't use `library()` or `require()`. These modify the search path, affecting what functions are available from the global environment. Instead, you should use the *DESCRIPTION* to specify your package's requirements, as described in Chapter 9. This also makes sure those packages are installed when your package is installed.

- Never use `source()` to load code from a file. `source()` modifies the current environment, inserting the results of executing the code. There is no reason to use `source()` inside your package, i.e., in a file below *R/*. Sometimes people `source()` files below *R/* during package development, but as we've explained in "Test Drive with load_all()" on page 58 and "Fast Feedback via load_all()" on page 85, `load_all()` is a much better way to load your current code for exploration. If you're using `source()` to create a dataset, it is better to use the methods in Chapter 7 for including data in a package.

Here is a nonexhaustive list of other functions that should be used with caution:

- `options()`
- `par()`
- `setwd()`
- `Sys.setenv()`
- `Sys.setlocale()`
- `set.seed()` (or anything that changes the state of the random number generator)

If you must use them, make sure to clean up after yourself. In the following section, we show how to do this using functions from the withr package and in base R.

The flip side of this coin is that you should avoid relying on the user's landscape, which might be different from yours. For example, functions that rely on sorting strings are dangerous, because sort order depends on the system locale. In the following code, we see that locales one might actually encounter in practice (C, English, French, etc.) differ in how they sort non-ASCII strings or uppercase versus lowercase letters:

```
x <- c("bernard", "bérénice", "béatrice", "boris")

withr::with_locale(c(LC_COLLATE = "fr_FR"), sort(x))
#> [1] "béatrice" "bérénice" "bernard"  "boris"
withr::with_locale(c(LC_COLLATE = "C"), sort(x))
#> [1] "bernard"  "boris"    "béatrice" "bérénice"

x <- c("a", "A", "B", "b", "A", "b")

withr::with_locale(c(LC_COLLATE = "en_CA"), sort(x))
#> [1] "a" "A" "A" "b" "b" "B"
withr::with_locale(c(LC_COLLATE = "C"), sort(x))
#> [1] "A" "A" "B" "a" "b" "b"
```

If you write your functions as if all users have the same system locale as you, your code might fail.

Manage State with withr

If you need to modify the R landscape inside a function, then it is important to ensure your change is reversed *on exit* of that function. This is exactly what `base::on.exit()` is designed to do. You use `on.exit()` inside a function to register code to run later, that restores the landscape to its original state. It is important to note that proper tools, such as `on.exit()`, work even if we exit the function abnormally, i.e., due to an error. This is why it's worth using the official methods described here over any do-it-yourself solution.

We usually manage state using the withr package (*https://withr.r-lib.org*), which provides a flexible, `on.exit()`-like toolkit (`on.exit()` itself is covered in the next section). `withr::defer()` can be used as a drop-in replacement for `on.exit()`. Why do we like withr so much? First, it offers many prebuilt convenience functions for state changes that come up often. We also appreciate withr's default stack-like behavior (LIFO = last in, first out), its usability in interactive sessions, and its `envir` argument (in more advanced usage).

The general pattern is to capture the original state, schedule its eventual restoration "on exit," then make the state change. Some setters, such as `options()` or `par()`, return the old value when you provide a new value, leading to usage that looks like this:

```
f <- function(x, y, z) {
  ...                        # width option "as found"
  old <- options(width = 20) # width option is 20
  defer(options(old))        # width option is 20
  ...                        # width option is 20
}                            # original width option restored
```

Certain state changes, such as modifying session options, come up so often that withr offers premade helpers. Table 6-2 shows a few of the state change helpers in withr that you are most likely to find useful.

Table 6-2. Selected functions from withr

Do / undo this	withr functions
Set an R option	`with_options()`, `local_options()`
Set an environment variable	`with_envvar()`, `local_envvar()`
Change working directory	`with_dir()`, `local_dir()`
Set a graphics parameter	`with_par()`, `local_par()`

You'll notice each helper comes in two forms that are useful in different situations:

- `with_*()` functions are best for executing small snippets of code with a temporarily modified state. (These functions are inspired by how `base::with()` works.)

  ```
  f <- function(x, sig_digits) {
    # imagine lots of code here
    withr::with_options(
      list(digits = sig_digits),
      print(x)
    )
    # ... and a lot more code here
  }
  ```

- `local_*()` functions are best for modifying state "from now until the function exits":

  ```
  g <- function(x, sig_digits) {
    withr::local_options(list(digits = sig_digits))
    print(x)
    # imagine lots of code here
  }
  ```

Developing code interactively with withr is pleasant, because deferred actions can be scheduled even on the global environment. Those cleanup actions can then be executed with `withr::deferred_run()` or cleared without execution with `withr::deferred_clear()`. Without this feature, it can be tricky to experiment with code that needs cleanup "on exit," because it behaves so differently when executed in the console versus at arm's length inside a function.

More in-depth coverage is given in the withr vignette Changing and restoring state (*https://oreil.ly/OQPuO*) and withr will also prove useful when we talk about testing in Chapter 13.

Restore State with base::on.exit()

Here is how the general "save, schedule restoration, change" pattern looks when using `base::on.exit()`:

```
f <- function(x, y, z) {
  ...
  old <- options(mfrow = c(2, 2), pty = "s")
  on.exit(options(old), add = TRUE)
  ...
}
```

Other state changes aren't available with that sort of setter and you must implement it yourself:

```
g <- function(a, b, c) {
  ...
  scratch_file <- tempfile()
  on.exit(unlink(scratch_file), add = TRUE)
  file.create(scratch_file)
  ...
}
```

Note that we specify `on.exit(..., add = TRUE)`, because you almost always want this behavior; i.e., to *add* to the list of deferred cleanup tasks rather than to *replace* them entirely. This (and the default value of `after`) are related to our preference for `withr::defer()`, when we're willing to take a dependency on withr. These issues are explored in a withr vignette (*https://oreil.ly/eDpcF*).

Isolate Side Effects

Creating plots and printing output to the console are two other ways of affecting the global R environment. Often you can't avoid these (because they're important!) but it's good practice to isolate them in functions that *only* produce output. This also makes it easier for other people to repurpose your work for new uses. For example, if you separate data preparation and plotting into two functions, others can use your data prep work (which is often the hardest part!) to create new visualizations.

When You Do Need Side Effects

Occasionally, packages do need side effects. This is most common if your package talks to an external system—you might need to do some initial setup when the package loads. To do that, you can use two special functions: `.onLoad()` and `.onAttach()`. These are called when the package is loaded and attached. You'll learn about the distinction between the two in "Attaching Versus Loading" on page 151. For now, you should always use `.onLoad()` unless explicitly directed otherwise.

Some common uses of `.onLoad()` and `.onAttach()` are:

- To set custom options for your package with `options()`. To avoid conflicts with other packages, ensure that you prefix option names with the name of your package. Also be careful not to override options that the user has already set. Here's a (highly redacted) version of dplyr's `.onLoad()` function, which sets an option that controls progress reporting:

  ```
  .onLoad <- function(libname, pkgname) {
    op <- options()
    op.dplyr <- list(
      dplyr.show_progress = TRUE
  ```

```
      )
      toset <- !(names(op.dplyr) %in% names(op))
      if (any(toset)) options(op.dplyr[toset])

      invisible()
    }
```

This allows functions in dplyr to use `getOption("dplyr.show_progress")` to determine whether to show progress bars, relying on the fact that a sensible default value has already been set.

- To display an informative message when the package is attached. This might make usage conditions clear or display package capabilities based on current system conditions. Startup messages are one place where you should use `.onAttach()` instead of `.onLoad()`. To display startup messages, always use `packageStartupMessage()`, and not `message()`. (This allows `suppress PackageStartupMessages()` to selectively suppress package startup messages).

```
    .onAttach <- function(libname, pkgname) {
      packageStartupMessage("Welcome to my package")
    }
```

As you can see in the examples, `.onLoad()` and `.onAttach()` are called with two arguments: `libname` and `pkgname`. They're rarely used (they're a holdover from the days when you needed to use `library.dynam()` to load compiled code). They give the path where the package is installed (the "library") and the name of the package.

If you use `.onLoad()`, consider using `.onUnload()` to clean up any side effects. By convention, `.onLoad()` and friends are usually saved in a file called *R/zzz.R*. (Note that `.First.lib()` and `.Last.lib()` are old versions of `.onLoad()` and `.onUnload()` and should no longer be used.)

One especially hairy thing to do in a function like `.onLoad()` or `.onAttach()` is to change the state of the random number generator. Once upon a time, ggplot2 used `sample()` when deciding whether to show a startup message, but only in interactive sessions. This, in turn, created a reproducibility puzzle for users who were using `set.seed()` for their own purposes, prior to attaching ggplot2 with `library(ggplot2)`, and running the code both interactively and noninteractively. The chosen solution was to wrap the offending startup code inside `withr::with_preserve_seed()`, which leaves the user's random seed as it found it.

Constant Health Checks

Here is a typical sequence of calls when using devtools for package development:

1. Edit one or more files below *R/*.
2. `document()` (if you've made any changes that impact help files or *NAMESPACE*).
3. `load_all()`.
4. Run some examples interactively.
5. `test()` (or `test_active_file()`).
6. `check()`.

An interesting question is how frequently and rapidly you move through this development cycle. We often find ourselves running through the preceding sequence several times in an hour or in a day while adding or modifying a single function.

Those newer to package development might be most comfortable slinging R code and much less comfortable writing and compiling documentation, simulating package build and installation, testing, and running `R CMD check`. And it is human nature to embrace the familiar and postpone the unfamiliar. This often leads to a dysfunctional workflow where the full sequence unfolds infrequently, maybe once per month or every couple of months, very slowly and often with great pain:

1. Edit one or more files below *R/*.
2. Build, install, and use the package. Iterate occasionally with previous step.
3. Write documentation (once the code is "done").
4. Write tests (once the code is "done").
5. Run `R CMD check` right before submitting to CRAN or releasing in some other way.

We've already talked about the value of fast feedback, in the context of `load_all()`. But this also applies to running `document()`, `test()`, and `check()`. There are defects you just can't detect from using `load_all()` and running a few interactive examples that are immediately revealed by more formal checks. Finding and fixing five bugs, one at a time, right after you created each one is much easier than troubleshooting all five at once (possibly interacting with each other), weeks or months after you last touched the code.

Submitting to CRAN

If you're planning on submitting your package to CRAN, you must use only ASCII characters in your *.R* files. In practice, this means you are limited to the digits 0 to 9, lowercase letters "a" to "z," uppercase letters "A" to "Z," and common punctuation.

But sometimes you need to inline a small bit of character data that includes, e.g., a Greek letter (μ), an accented character (ü), or a symbol (30°). You can use any Unicode character as long as you specify it in the special Unicode escape "\u1234" format. The easiest way to find the correct code point is to use stringi::stri_escape_unicode():

```
x <- "This is a bullet •"
y <- "This is a bullet \u2022"
identical(x, y)
#> [1] TRUE
cat(stringi::stri_escape_unicode(x))
#> This is a bullet \u2022
```

Sometimes you have the opposite problem. You don't *intentionally* have any non-ASCII characters in your R code, but automated checks reveal that you do:

```
W  checking R files for non-ASCII characters ...
   Found the following file with non-ASCII characters:
     foo.R
   Portable packages must use only ASCII characters in
   their R code, except perhaps in comments.
   Use \uxxxx escapes for other characters.
```

The most common offenders are "curly" or "smart" single and double quotes that sneak in through copy/paste. The functions tools::showNonASCII() and tools::showNonASCIIfile(file) help you find the offending file(s) and line(s):

```
tools::showNonASCIIfile("R/foo.R")
#> 666: #' If you<e2><80><99>ve copy/pasted quotes,
#> watch out!
```

## CHAPTER 7

Data

It's often useful to include data in a package. If the primary purpose of a package is to distribute useful functions, example datasets make it easier to write excellent documentation. These datasets can be handcrafted to provide compelling use cases for the functions in the package. Here are some examples of this type of package data:

tidyr
> `billboard` (song rankings), `who` (tuberculosis data from the World Health Organization)

dplyr
> `starwars` (Star Wars characters), `storms` (storm tracks)

At the other extreme, some packages exist solely for the purpose of distributing data, along with its documentation. These are sometimes called "data packages." A data package can be a nice way to share example data across multiple packages. It is also a useful technique for getting relatively large, static files out of a more function-oriented package, which might require more frequent updates. Here are some examples of data packages:

- nycflights13 (*https://nycflights13.tidyverse.org*)
- babynames (*http://hadley.github.io/babynames*)

Finally, many packages benefit from having internal data that is used for internal purposes, but that is not directly exposed to the users of the package.

In this chapter we describe useful mechanisms for including data in your package. The practical details differ depending on who needs access to the data, how often it changes, and what they will do with it:

- If you want to store R objects and make them available to the user, put them in *data/*. This is the best place to put example datasets. All the concrete examples we've given so far for data in a package and data as a package use this mechanism. See section "Exported Data" on page 100.

- If you want to store R objects for your own use as a developer, put them in *R/sysdata.rda*. This is the best place to put internal data that your functions need. See "Internal Data" on page 105.

- If you want to store data in some raw, non-R-specific form and make it available to the user, put it in *inst/extdata/*. For example, readr and readxl each use this mechanism to provide a collection of delimited files and Excel workbooks, respectively. See "Raw Data File" on page 106.

- If you want to store dynamic data that reflects the internal state of your package within a single R session, use an environment. This technique is not as common or well-known as the previous ones but can be very useful in specific situations. See "Internal State" on page 109.

- If you want to store data persistently across R sessions, such as configuration or user-specific data, use one of the officially sanctioned locations. See "Persistent User Data" on page 112.

Exported Data

The most common location for package data is (surprise!) *data/*. We recommend that each file in this directory be an *.rda* file created by `save()` containing a single R object, with the same name as the file. The easiest way to achieve this is to use `usethis::use_data()`:

```
my_pkg_data <- sample(1000)
usethis::use_data(my_pkg_data)
```

Let's imagine we are working on a package named "pkg." The preceding snippet creates *data/my_pkg_data.rda* inside the source of the pkg package and adds `Lazy Data: true` in your DESCRIPTION. This makes the `my_pkg_data` R object available to users of pkg via `pkg::my_pkg_data` or, after attaching pkg with `library(pkg)`, as `my_pkg_data`.

The preceding snippet is something the maintainer executes once (or every time they need to update my_pkg_data). This is workflow code and should *not* appear in the *R/* directory of the source package. (We'll talk about a suitable place to keep this code shortly.) For larger datasets, you may want to experiment with the compression setting, which is under the control of the compress argument. The default is "bzip2," but sometimes "gzip" or "xz" can create smaller files.

It's possible to use other types of files below *data/*, but we don't recommend it because *.rda* files are already fast, small, and explicit. The other possibilities are described in the documentation for utils::data() and in the "Data in packages" (*https://oreil.ly/_X0fI*) section of *Writing R Extensions*. In terms of advice to package authors, the help topic for data() seems to implicitly make the same recommendations as we do:

- Store one R object in each *data/*.rda* file.
- Use the same name for that object and its *.rda* file.
- Use lazy-loading, by default.

If the DESCRIPTION contains LazyData: true, then datasets will be lazily loaded. This means that they won't occupy any memory until you use them. The following example shows memory usage before and after loading the nycflights13 package. You can see that memory usage doesn't change significantly until you inspect the flights dataset stored inside the package:

```
lobstr::mem_used()
#> 56.81 MB
library(nycflights13)
lobstr::mem_used()
#> 58.73 MB

invisible(flights)
lobstr::mem_used()
#> 99.43 MB
```

We recommend that you include LazyData: true in your DESCRIPTION if you are shipping *.rda* files below *data/*. If you use use_data() to create such datasets, it will automatically make this modification to DESCRIPTION for you.

It is important to note that lazily loaded datasets do *not* need to be preloaded with utils::data() and, in fact, it's usually best to avoid doing so. In the preceding example, once we did library(nyc flights13), we could immediately access flights. There is no call to data(flights), because it is not necessary.

There are specific downsides to data(some_pkg_data) calls that support a policy of using data() only when it is actually necessary, i.e., for datasets that would not be available otherwise:

- By default, data(some_pkg_data) creates one or more objects in the user's global workspace. There is the potential to silently overwrite preexisting objects with new values.

- There is also no guarantee that data(foo) will create exactly one object named "foo." It could create more than one object and/or objects with totally different names.

One argument in favor of calls like data(some_pkg_data, package = "pkg") that are not strictly necessary is that it clarifies which package provides some_pkg_data. We prefer alternatives that don't modify the global workspace, such as a code comment or access via pkg::some_pkg_data.

This excerpt from the documentation of data() conveys that it is largely of historical importance:

> data() was originally intended to allow users to load datasets from packages for use in their examples, and as such it loaded the datasets into the workspace .Global Env. This avoided having large datasets in memory when not in use: that need has been almost entirely superseded by lazy-loading of datasets.

Preserve the Origin Story of Package Data

Often, the data you include in *data/* is a cleaned-up version of raw data you've gathered from elsewhere. We highly recommend taking the time to include the code used to do this in the source version of your package. This makes it easy for you to update or reproduce your version of the data. This data-creating script is also a natural place to leave comments about important properties of the data, i.e., which features are important for downstream usage in package documentation.

We suggest that you keep this code in one or more *.R* files below *data-raw/*. You don't want it in the bundled version of your package, so this folder should be listed in *.Rbuildignore*. usethis has a convenience function that can be called when you first adopt the *data-raw/* practice or when you add an additional *.R* file to the folder:

```
usethis::use_data_raw()
```

```
usethis::use_data_raw("my_pkg_data")
```

use_data_raw() creates the *data-raw/* folder and lists it in *.Rbuildignore*. A typical script in *data-raw/* includes code to prepare a dataset and ends with a call to use_data().

These data packages all use the approach recommended here for *data-raw/*:

- babynames (*https://github.com/hadley/babynames*)
- nycflights13 (*https://github.com/hadley/nycflights13*)
- gapminder (*https://github.com/jennybc/gapminder*)

ggplot2: A Cautionary Tale

We have a confession to make: the origins of many of ggplot2's example datasets have been lost in the sands of time. In the grand scheme of things, this is not a huge problem, but maintenance is certainly more pleasant when a package's assets can be reconstructed *de novo* and easily updated as necessary.

Submitting to CRAN

Generally, package data should be smaller than a megabyte—if it's larger you'll need to argue for an exemption. This is usually easier to do if the data is in its own package and won't be updated frequently, i.e., if you approach this as a dedicated "data package." For reference, the babynames and nycflights packages have had a release once every one to two years since they first appeared on CRAN.

If you are bumping up against size issues, you should be intentional about the method of data compression. The default for usethis::use_data(compress =) is "bzip2" whereas the default for save(compress =) is (effectively) "gzip," and "xz" is yet another valid option.

You'll have to experiment with different compression methods and make this decision empirically. tools::resaveRdaFiles("data/") automates this process but doesn't inform you of which compression method was chosen. You can learn this after the fact with tools::checkRdaFiles(). Assuming you are keeping track of the code to generate your data, it would be wise to update the corresponding use_data(compress =) call below *data-raw/* and regenerate the *.rda* cleanly.

Documenting Datasets

Objects in *data/* are always effectively exported (they use a slightly different mechanism than *NAMESPACE* but the details are not important). This means that they must be documented. Documenting data is like documenting a function with a few minor differences. Instead of documenting the data directly, you document the name of the dataset and save it in *R/*. For example, the roxygen2 block used to document the who data in tidyr is saved in *R/data.R* and looks something like this:

```
#' World Health Organization TB data
#'
#' A subset of data from the World Health Organization Global Tuberculosis
#' Report ...
#'
#' @format ## `who`
#' A data frame with 7,240 rows and 60 columns:
#' \describe{
#'   \item{country}{Country name}
#'   \item{iso2, iso3}{2 & 3 letter ISO country codes}
#'   \item{year}{Year}
#'   ...
#' }
#' @source <https://www.who.int/teams/global-tuberculosis-programme/data>
"who"
```

Two roxygen tags are especially important for documenting datasets:

- @format gives an overview of the dataset. For data frames, you should include a definition list that describes each variable. It's usually a good idea to describe variables' units here.

- @source provides details of where you got the data, often a URL.

Never @export a dataset.

Non-ASCII Characters in Data

The R objects you store in *data/*.rda* often contain strings, with the most common example being character columns in a data frame. If you can constrain these strings to only use ASCII characters, it certainly makes things simpler. But of course, there are plenty of legitimate reasons why package data might include non-ASCII characters.

In that case, we recommend that you embrace the UTF-8 Everywhere manifesto (*http://utf8everywhere.org*) and use the UTF-8 encoding. The *DESCRIPTION* file placed by usethis::create_package() always includes Encoding: UTF-8, so by default a devtools-produced package already advertises that it will use UTF-8.

Making sure that the strings embedded in your package data have the intended encoding is something you accomplish in your data preparation code, i.e., in the R scripts below *data-raw/*. You can use `Encoding()` to learn the current encoding of the elements in a character vector and functions such as `enc2utf8()` or `iconv()` to convert between encodings.

Submitting to CRAN

If you have UTF-8-encoded strings in your package data, you may see this from `R CMD check`:

```
- checking data for non-ASCII characters ... NOTE
  Note: found 352 marked UTF-8 strings
```

This `NOTE` is truly informational. It requires no action from you. As long as you actually intend to have UTF-8 strings in your package data, all is well.

Ironically, this `NOTE` is actually suppressed by `R CMD check --as-cran`, despite the fact that this NOTE does appear in the check results once a package is on CRAN (which implies that CRAN does not necessarily check with `--as-cran`). By default, `devtools::check()` sets the `--as-cran` flag and therefore does not transmit this `NOTE`. But you can surface it with `check(cran = FALSE, env_vars = c("`*R_CHECK_PACKAGE_DATA* *SETS_SUPPRESS_NOTES*`" = "false"))`.

Internal Data

Sometimes your package functions need access to precomputed data. If you put these objects in *data/*, they'll also be available to package users, which is not appropriate. Sometimes the objects you need are small and simple enough that you can define them with `c()` or `data.frame()` in the code below *R/*, perhaps in *R/data.R*. Larger or more complicated objects should be stored in your package's internal data in *R/sysdata.rda*.

Here are some examples of internal package data:

- Two color-related packages, munsell (*https://oreil.ly/SWNuh*) and dichromat (*https://oreil.ly/brhL8*), use *R/sysdata.rda* to store large tables of color data.

- googledrive (*https://oreil.ly/e3UsU*) and googlesheets4 (*https://oreil.ly/66Key*) wrap the Google Drive and Google Sheets APIs, respectively. Both use *R/sysdata.rda* to store data derived from a so-called Discovery Document (*https://oreil.ly/k1cxc*), which "describes the surface of the API, how to access the API and how API requests and responses are structured."

The easiest way to create *R/sysdata.rda* is to use `usethis::use_data(internal = TRUE)`:

```
internal_this <- ...
internal_that <- ...

usethis::use_data(internal_this, internal_that, internal = TRUE)
```

Unlike *data/*, where you use one *.rda* file per exported data object, you store all of your internal data objects together in the single file *R/sysdata.rda*.

Let's imagine we are working on a package named "pkg." The preceding snippet creates *R/sysdata.rda* inside the source of the pkg package. This makes the objects `internal_this` and `internal_that` available for use inside of the functions defined below *R/* and in the tests. During interactive development, `internal_this` and `internal_that` are available after a call to `devtools::load_all()`, just like an internal function.

Much of the advice given for external data holds for internal data as well:

- It's a good idea to store the code that generates your individual internal data objects, as well as the `use_data()` call that writes all of them into *R/sysdata.rda*. This is workflow code that belongs below *data-raw/*, not below *R/*.

- `usethis::use_data_raw()` can be used to initiate the use of *data-raw/* or to initiate a new `.R` script there.

- If your package is uncomfortably large, experiment with different values of `compress` in `use_data(internal = TRUE)`.

There are also key distinctions, where the handling of internal and external data differs:

- Objects in *R/sysdata.rda* are not exported (they shouldn't be), so they don't need to be documented.

- Usage of *R/sysdata.rda* has no impact on *DESCRIPTION*, i.e., the need to specify the `LazyData` field is strictly about the exported data below *data/*.

Raw Data File

If you want to show examples of loading/parsing raw data, put the original files in *inst/extdata/*. When the package is installed, all files (and folders) in *inst/* are moved up one level to the top-level directory, which is why they can't have names that conflict with standard parts of an R package, like *R/* or *DESCRIPTION*. The files below *inst/extdata/* in the source package will be located below *extdata/* in the corresponding installed package. You may want to revisit Figure 3-1 to review the file structure for different package states.

The main reason to include such files is when a key part of a package's functionality is to act on an external file. Examples of such packages include:

- readr, which reads rectangular data out of delimited files
- readxl, which reads rectangular data out of Excel spreadsheets
- xml2, which can read XML and HTML from file
- archive, which can read archive files, such as tar or ZIP

All of these packages have one or more example files below *inst/extdata/*, which are useful for writing documentation and tests.

It is also common for data packages to provide, e.g., a *CSV* version of the package data that is also provided as an R object. Examples of such packages include:

palmerpenguins
> penguins and penguins_raw are also represented as *extdata/penguins.csv* and *extdata/penguins_raw.csv*

gapminder
> gapminder, continent_colors, and country_colors are also represented as *extdata/gapminder.tsv*, *extdata/continent-colors.tsv*, and *extdata/country-colors.tsv*

This has two payoffs: First, it gives teachers and other expositors more to work with once they decide to use a specific dataset. If you've started teaching R with palmerpenguins::penguins or gapminder::gapminder and you want to introduce data import, it can be helpful to students if their first use of a new command, like readr::read_csv() or read.csv(), is applied to a familiar dataset. They have preexisting intuition about the expected result. Finally, if package data evolves over time, having a *CSV* or other plain-text representation in the source package can make it easier to see what's changed.

Filepaths

The path to a package file found below *extdata/* clearly depends on the local environment, i.e., it depends on where installed packages live on that machine. The base function system.file() can report the full path to files distributed with an R package. It can also be useful to *list* the files distributed with an R package:

```
system.file("extdata", package = "readxl") |> list.files()
#> [1] "clippy.xls"    "clippy.xlsx"   "datasets.xls"  "datasets.xlsx"
#> [5] "deaths.xls"    "deaths.xlsx"   "geometry.xls"  "geometry.xlsx"
#> [9] "type-me.xls"   "type-me.xlsx"

system.file("extdata", "clippy.xlsx", package = "readxl")
#> [1] "/Users/jenny/Library/R/x86_64/4.2/library/readxl/extdata/clippy.xlsx"
```

These filepaths present yet another workflow dilemma: when you're developing your package, you engage with it in its source form, but your users engage with it as an installed package. Happily, devtools provides a shim for base::system.file() that is activated by load_all(). This makes interactive calls to system.file() from the global environment and calls from within the package namespace "just work."

Be aware that, by default, system.file() returns the empty string, not an error, for a file that does not exist:

```
system.file("extdata", "I_do_not_exist.csv", package = "readr")
#> [1] ""
```

If you want to force a failure in this case, specify mustWork = TRUE:

```
system.file("extdata", "I_do_not_exist.csv", package = "readr", mustWork = TRUE)
#> Error in system.file("extdata", "I_do_not_exist.csv", package = "readr",
#> : no file found
```

The fs package (*https://fs.r-lib.org*) offers fs::path_package(). This is essentially base::system.file() with a few added features that we find advantageous, whenever it's reasonable to take a dependency on fs:

- It errors if the filepath does not exist.

- It throws distinct errors when the package does not exist versus when the file does not exist within the package.

- During development, it works for interactive calls, calls from within the loaded package's namespace, and even for calls originating in dependencies:

```
fs::path_package("extdata", package = "idonotexist")
#> Error: Can't find package `idonotexist` in library locations:
#>   - '/Users/jenny/Library/R/x86_64/4.2/library'
#>   - '/Library/Frameworks/R.framework/Versions/4.2/Resources/library'

fs::path_package("extdata", "I_do_not_exist.csv", package = "readr")
#> Error: File(s) '/Users/jenny/.../extdata/I_do_not_exist.csv' do not exist

fs::path_package("extdata", "chickens.csv", package = "readr")
#> /Users/jenny/Library/R/x86_64/4.2/library/readr/extdata/chickens.csv
```

pkg_example() Path Helpers

We like to offer convenience functions that make example files easy to access. These are just user-friendly wrappers around `system.file()` or `fs::path_package()`, but they can have added features, such as the ability to list the example files. Here's the definition and some usage of `readxl::readxl_example()`:

```
readxl_example <- function(path = NULL) {
  if (is.null(path)) {
    dir(system.file("extdata", package = "readxl"))
  } else {
    system.file("extdata", path, package = "readxl", mustWork = TRUE)
  }
}

readxl::readxl_example()
#>  [1] "clippy.xls"    "clippy.xlsx"    "datasets.xls"   "datasets.xlsx"
#>  [5] "deaths.xls"    "deaths.xlsx"    "geometry.xls"   "geometry.xlsx"
#>  [9] "type-me.xls"   "type-me.xlsx"

readxl::readxl_example("clippy.xlsx")
#> [1] "/Users/jenny/Library/R/x86_64/4.2/library/readxl/extdata/clippy.xlsx"
```

Internal State

Sometimes there's information that multiple functions from your package need to access that:

- Must be determined at load time (or even later), not at build time. It might even be dynamic.

- Doesn't make sense to pass in via a function argument. Often it's some obscure detail that a user shouldn't even know about.

A great way to manage such data is to use an *environment*.[1] This environment must be created at build time, but you can populate it with values after the package has been loaded and update those values over the course of an R session. This works because environments have reference semantics (whereas more pedestrian R objects, such as atomic vectors, lists, or data frames have value semantics).

---

[1] If you don't know much about R environments and what makes them special, a great resource is the Environments chapter (*https://adv-r.hadley.nz/environments.html*) of *Advanced R*.

Consider a package that can store the user's favorite letters or numbers. You might start out with code like this in a file below *R/*:

```
favorite_letters <- letters[1:3]

#' Report my favorite letters
#' @export
mfl <- function() {
  favorite_letters
}

#' Change my favorite letters
#' @export
set_mfl <- function(l = letters[24:26]) {
  old <- favorite_letters
  favorite_letters <<- l
  invisible(old)
}
```

favorite_letters is initialized to ("a", "b", "c") when the package is built. The user can then inspect favorite_letters with mfl(), at which point they'll probably want to register *their* favorite letters with set_mfl(). Note that we've used the super assignment operator <<- in set_mfl() in the hope that this will reach up into the package environment and modify the internal data object favorite_letters. But a call to set_mfl() fails like so:[2]

```
mfl()
#> [1] "a" "b" "c"

set_mfl(c("j", "f", "b"))
#> Error in set_mfl() :
#>   cannot change value of locked binding for 'favorite_letters'
```

Because favorite_letters is a regular character vector, modification requires making a copy and rebinding the name favorite_letters to this new value. And that is what's disallowed: you can't change the binding for objects in the package namespace (well, at least not without trying harder than this). Defining favorite_letters this way only works if you will never need to modify it.

However, if we maintain state within an internal package environment, we *can* modify objects contained in the environment (and even add completely new objects). Here's an alternative implementation that uses an internal environment named "the":

```
the <- new.env(parent = emptyenv())
the$favorite_letters <- letters[1:3]
```

---

2 This example will execute without error if you define favorite_letters, mfl(), and set_mfl() in the global workspace and call set_mfl() in the console. But this code will fail once favorite_letters, mfl(), and set_mfl() are defined *inside a package*.

```
#' Report my favorite letters
#' @export
mfl2 <- function() {
  the$favorite_letters
}

#' Change my favorite letters
#' @export
set_mfl2 <- function(l = letters[24:26]) {
  old <- the$favorite_letters
  the$favorite_letters <- l
  invisible(old)
}
```

Now a user *can* register their favorite letters:

```
mfl2()
#> [1] "a" "b" "c"

set_mfl2(c("j", "f", "b"))

mfl2()
#> [1] "j" "f" "b"
```

Note that this new value for the$favorite_letters persists only for the remainder of the current R session (or until the user calls set_mfl2() again). More precisely, the altered state persists only until the next time the package is loaded (including via load_all()). At load time, the environment the is reset to an environment containing exactly one object, named favorite_letters, with value ("a", "b", "c"). It's like the movie *Groundhog Day*. (We'll discuss more persistent package- and user-specific data in the next section.)

Jim Hester introduced our group to the nifty idea of using "the" as the name of an internal package environment. This lets you refer to the objects inside in a very natural way, such as the$token, meaning "*the* token." It is also important to specify parent = emptyenv() when defining an internal environment, as you generally don't want the environment to inherit from any other (nonempty) environment.

As seen in the favorite letters example, the definition of the environment should happen as a top-level assignment in a file below *R/*. (In particular, this is a legitimate reason to define a nonfunction at the top-level of a package; see "Understand When Code Is Executed" on page 87 for why this should be rare.) As for where to place this definition, there are two considerations:

- Define it before you use it. If other top-level calls refer to the environment, the definition must come first when the package code is being executed at build time. This is why *R/aaa.R* is a common and safe choice.

- Make it easy to find later when you're working on related functionality. If an environment is only used by one family of functions, define it there. If environment usage is sprinkled around the package, define it in a file with package-wide connotations.

Here are some examples of how packages use an internal environment:

googledrive
> Various functions need to know the file ID for the current user's home directory on Google Drive. This requires an API call (a relatively expensive and error-prone operation), which yields an eye-watering string of ~40 seemingly random characters that only a computer can love. It would be inhumane to expect a user to know this or to pass it into every function. It would also be inefficient to rediscover the ID repeatedly. Instead, googledrive determines the ID upon first need, then caches it for later use.

usethis
> Most functions need to know the active project, i.e., which directory to target for file modification. This is often the current working directory, but that is not an invariant usethis can rely upon. One potential design is to make it possible to specify the target project as an argument of every function in usethis. But this would create significant clutter in the user interface, as well as internal fussiness. Instead, we determine the active project upon first need, cache it, and provide methods for (re)setting it.

The blog post Package-Wide Variables/Cache in R Packages (*https://oreil.ly/-8-zo*) gives a more detailed development of this technique.

Persistent User Data

Sometimes there is data that your package obtains, on behalf of itself or the user, that should persist *even across R sessions*. This is our last and probably least common form of storing package data. For the data to persist this way, it has to be stored on disk and the big question is where to write such a file.

This problem is hardly unique to R. Many applications need to leave notes to themselves. It is best to comply with external conventions, which in this case means the XDG Base Directory Specification (*https://oreil.ly/SC63H*). You need to use the official locations for persistent file storage, because it's the responsible and courteous thing to do and also to comply with CRAN policies.

Submitting to CRAN

You can't just write persistent data into the user's home directory. Here's a relevant excerpt from the CRAN policy at the time of writing:

> Packages should not write in the user's home filespace (including clipboards), nor anywhere else on the file system apart from the R session's temporary directory ....

> For R version 4.0 or later (hence a version dependency is required or only conditional use is possible), packages may store user-specific data, configuration and cache files in their respective user directories obtained from `tools::R_user_dir()`, provided that by [sic] default sizes are kept as small as possible and the contents are actively managed (including removing outdated material).

The primary function you should use to derive acceptable locations for user data is `tools::R_user_dir()`.[3] Here are some examples of the generated filepaths:

```
tools::R_user_dir("pkg", which = "data")
#> [1] "/Users/jenny/Library/Application Support/org.R-project.R/R/pkg"
tools::R_user_dir("pkg", which = "config")
#> [1] "/Users/jenny/Library/Preferences/org.R-project.R/R/pkg"
tools::R_user_dir("pkg", which = "cache")
#> [1] "/Users/jenny/Library/Caches/org.R-project.R/R/pkg"
```

One last thing you should consider with respect to persistent data is: does this data *really* need to persist? Do you *really* need to be the one responsible for storing it?

If the data is potentially sensitive, such as user credentials, it is recommended to obtain the user's consent to store it, i.e., to require interactive consent when initiating the cache. Also consider that the user's operating system or command-line tools might provide a means of secure storage that is superior to any DIY solution that you might implement. The packages keyring (*https://oreil.ly/9ZIiH*), gitcreds (*https://gitcreds.r-lib.org*), and credentials (*https://oreil.ly/7FaLZ*) are examples of packages that tap into externally provided tooling. Before embarking on any creative solution for storing secrets, consider that your effort is probably better spent integrating with an established tool.

---

3 Note that `tools::R_user_dir()` first appeared in R 4.0. If you need to support older versions of R, then you should use the rappdirs package (*https://rappdirs.r-lib.org*), which is a port of the Python appdirs module, and which follows the tidyverse policy regarding R version support (*https://oreil.ly/B5uUW*), meaning the minimum supported R version is advancing and will eventually slide past R 4.0. rappdirs produces different filepaths than `tools::R_user_dir()`. However, both tools implement something that is consistent with the XDG spec, just with different opinions about how to create filepaths beyond what the spec dictates.

## CHAPTER 8

Other Components

The first two chapters in this part of the book cover the two most obvious things that people distribute via an R package: functions (Chapter 6) and data (Chapter 7). But that's not all it takes to make an R package. There are other package components that are either required, such as a *DESCRIPTION* file, or highly recommended, such as tests and documentation.

The next few parts of the book are organized around important concepts: dependencies, testing, and documentation. But before we dig into those topics, this chapter demystifies some package parts that are not needed in every package, but that are nice to be aware of.

Other Directories

Here are some top-level directories you might encounter in an R source package, in rough order of importance and frequency of use:

src/

This directory contains source and header files for compiled code, most often C and C++. This is an important technique that is used to make R packages more performant and to unlock the power of external libraries for R users. As of the second edition, the book no longer covers this topic, since a truly useful treatment of compiled code requires more space than we can give it here. The tidyverse generally uses the cpp11 package (*https://cpp11.r-lib.org*) to connect C++ to R; most other packages use Rcpp (*https://www.rcpp.org*), the most well-established package for integrating R and C++.

inst

> This directory can hold arbitrary additional files that you want include in your package. This includes a few special files, like the *CITATION*, described in the next section. Other examples of files that might appear below *inst/* include R Markdown templates (see `usethis::use_rmarkdown_template()`) or RStudio add-ins (*https://oreil.ly/1keh5*).

tools/

> This directory can contain auxiliary files needed during configuration, usually found in the company of a `configure` script. We discuss this more in "Configuration Tools" on page 119.

demo/

> This directory is for package demos. We regard demos as a legacy phenomenon, whose goals are now better met by vignettes (Chapter 17). For actively maintained packages, it probably makes sense to repurpose the content in any existing demos somewhere that's more visible, e.g., in *README.Rmd* (see "README" on page 273) or in vignettes (see Chapter 17). These other locations offer other advantages, such as making sure that the code is exercised regularly. This is not true of actual demos, leaving them vulnerable to rot.

exec/

> This directory can contain executable scripts. Unlike files placed in other directories, files in *exec/* are automatically flagged as executable. Empirically, to the extent that R packages are shipping scripts for external interpreters, the *inst/* directory seems to be the preferred location these days.

po/

> This directory provides translations for messages. This is useful but beyond the scope of this book. See the "Internationalization" (*https://oreil.ly/5AJYr*) chapter of *Writing R Extensions* and the potools package (*https://oreil.ly/IL9y1*) for more details.

Installed Files

When a package is installed, everything in *inst/* is copied into the top-level directory of the installed package (see Figure 3-1). In some sense *inst/* is the opposite of *.Rbuildignore*; where *.Rbuildignore* lets you remove arbitrary files and directories from the built package, *inst/* lets you add them.

You are free to put anything you like in *inst/* with one caution: because *inst/* is copied into the top-level directory, don't create a subdirectory that collides with any of the directories that make up the official structure of an R package. We recommend avoiding directories with special significance in either the source or installed form of a package, such as *inst/data, inst/help, inst/html, inst/libs, inst/man, inst/Meta, inst/R, inst/src, inst/tests, inst/tools,* and *inst/vignettes*. In most cases, this prevents you from having a malformed package. And even though some of these directories are technically allowed, they can be an unnecessary source of confusion.

Here are some of the most common files and folders found in *inst/*:

inst/CITATION
    How to cite the package, see "Package Citation" for details.

inst/extdata
    Additional external data for examples and vignettes. See "Raw Data File" on page 106 for more details.

What if you need a path to the file at *inst/foo* to use in, e.g., the code below *R/* or in your documentation? The default solution is to use `system.file("foo", package = "yourpackage")`. But this presents a workflow dilemma: when you're developing your package, you engage with it in its source form (*inst/foo*), but your users engage with its installed form (*/foo*). Happily, devtools provides a shim for `system.file()` that is activated by `load_all()`. "Filepaths" on page 107 covers this in more depth and includes an interesting alternative, `fs::path_package()` .

Package Citation

The *CITATION* file lives in the *inst* directory and is intimately connected to the `citation()` function, which tells you how to cite R and R packages. Calling `cita tion()` without any arguments tells you how to cite base R:

```
citation()
#>
#> To cite R in publications use:
#>
#>   R Core Team (2022). R: A language and environment for
#>   statistical computing. R Foundation for Statistical
#>   Computing, Vienna, Austria. URL
#>   https://www.R-project.org/.
#>
#> A BibTeX entry for LaTeX users is
#>
#>   @Manual{,
```

```
#>      title = {R: A Language and Environment for Statistical Computing},
#>      author = {{R Core Team}},
#>      organization = {R Foundation for Statistical Computing},
#>      address = {Vienna, Austria},
#>      year = {2022},
#>      url = {https://www.R-project.org/},
#>    }
#>
#> We have invested a lot of time and effort in creating R,
#> please cite it when using it for data analysis. See also
#> 'citation("pkgname")' for citing R packages.
```

Calling it with a package name tells you how to cite that package:

```
citation("tidyverse")
#>
#> To cite package 'tidyverse' in publications use:
#>
#>   Wickham H, Averick M, Bryan J, Chang W, McGowan LD,
#>   François R, Grolemund G, Hayes A, Henry L, Hester J, Kuhn
#>   M, Pedersen TL, Miller E, Bache SM, Müller K, Ooms J,
#>   Robinson D, Seidel DP, Spinu V, Takahashi K, Vaughan D,
#>   Wilke C, Woo K, Yutani H (2019). "Welcome to the
#>   tidyverse." _Journal of Open Source Software_, 4(43),
#>   1686. doi:10.21105/joss.01686
#>   <https://doi.org/10.21105/joss.01686>.
#>
#> A BibTeX entry for LaTeX users is
#>
#>   @Article{,
#>     title = {Welcome to the {tidyverse}},
#>     author = {Hadley Wickham and Mara Averick and Jennifer Bryan and...},
#>     year = {2019},
#>     journal = {Journal of Open Source Software},
#>     volume = {4},
#>     number = {43},
#>     pages = {1686},
#>     doi = {10.21105/joss.01686},
#>   }
```

The associated *inst/CITATION* file looks like this:

```
bibentry(
  "Article",
  title = "Welcome to the {tidyverse}",
  author = "Hadley Wickham, Mara Averick, Jennifer Bryan, Winston Chang...,
  year = 2019,
  journal = "Journal of Open Source Software",
  volume = 4,
  number = 43,
  pages = 1686,
  doi = "10.21105/joss.01686",
)
```

You can call `usethis::use_citation()` to initiate this file and fill in your details. Read the `?bibentry` help topic for more details.

Configuration Tools

If a package has a configuration script (`configure` on Unix-alikes, `configure.win` on Windows), it is executed as the first step by `R CMD INSTALL`. This is typically associated with a package that has an *src/* subdirectory containing C/C++ code and the `configure` script is needed at compile time. If that script needs auxiliary files, those should be located in the *tools/* directory. The scripts below *tools/* can have an effect on the installed package, but the contents of *tools/* will not ultimately be present in the installed package. In any case, this is mostly (but not solely) relevant to packages with compiled code, which is beyond the scope of this book.

We bring this up because, in practice, some packages use the *tools/* directory for a different but related purpose. Some packages have periodic maintenance tasks for which it is helpful to record detailed instructions. For example, many packages embed some sort of external resource, e.g., code or data:

- Source code and headers for an embedded third-party C/C++ library
- Web toolkits
- R code that's inlined (as opposed to imported)
- Specification for a web API
- Color palettes, styles, and themes

These external assets are also usually evolving over time, so they need to be re-ingested on a regular basis. This makes it particularly rewarding to implement such housekeeping programmatically.

This is the second, unofficial use of the *tools/* directory, characterized by two big differences with its official purpose: the packages that do this generally do not have a `configure` script and they list *tools/* in *.Rbuildignore*, meaning that these scripts are not included in the package bundle. These scripts are maintained in the source package for developer convenience but are never shipped with the package.

This practice is closely related to our recommendation to store the instructions for the creation of package data in *data-raw/* (see "Preserve the Origin Story of Package Data" on page 102) and to record the method of construction for any test fixtures (see "Store a Concrete useful_thing Persistently" on page 221).

## PART III

Package Metadata

## CHAPTER 9

DESCRIPTION

DESCRIPTION and *NAMESPACE* are two important files that provide metadata about your package. The *DESCRIPTION* file provides overall metadata about the package, such as the package name and which other packages it depends on. The *NAMESPACE* file specifies which functions your package makes available for others to use and, optionally, imports functions from other packages.

In this chapter, you'll learn about the most important fields found in *DESCRIPTION*. The next two chapters cover the topic of package dependencies, which is where the importance of the *NAMESPACE* file will become clear. First, in Chapter 10, we discuss the costs and benefits of dependencies and also provide the relevant technical context around how R finds objects. In Chapter 11, we explain the practical moves necessary to use your dependencies within your package. The metadata part of the book concludes with Chapter 12, which covers licensing.

The DESCRIPTION File

The job of the *DESCRIPTION* file is to store important metadata about your package. When you first start writing packages, you'll mostly use this metadata to record what packages are needed to run your package. However, as time goes by, other aspects of the metadata file will become useful to you, such as revealing what your package does (via the `Title` and `DESCRIPTION`) and whom to contact (you!) if there are any problems.

Every package must have a *DESCRIPTION*. In fact, it's the defining feature of a package (RStudio and devtools consider any directory containing *DESCRIPTION* to be a package).[1] To get you started, usethis::create_package("mypackage") automatically adds a bare-bones *DESCRIPTION* file. This will allow you to start writing the package without having to worry about the metadata until you need to. This minimal *DESCRIPTION* will vary a bit depending on your settings, but should look something like this:

```
Package: mypackage
Title: What the Package Does (One Line, Title Case)
Version: 0.0.0.9000
Authors@R:
    person("First", "Last", , "first.last@example.com", role = c("aut", "cre"),
           comment = c(ORCID = "YOUR-ORCID-ID"))
Description: What the package does (one paragraph).
License: `use_mit_license()`, `use_gpl3_license()` or friends to pick a
    license
Encoding: UTF-8
Roxygen: list(markdown = TRUE)
RoxygenNote: 7.2.3
```

If you create a lot of packages, you can customize the default content of new *DESCRIPTION* files by setting the global option usethis.description to a named list. You can preconfigure your preferred name, email, license, etc. See the article on usethis setup (*https://usethis.r-lib.org/articles/articles/usethis-setup.html*) for more details.

DESCRIPTION uses a simple file format called DCF, the Debian control format. You can see most of the structure in the examples in this chapter. Each line consists of a *field* name and a value, separated by a colon. When values span multiple lines, they need to be indented:

```
Description: The description of a package usually spans multiple lines.
    The second and subsequent lines should be indented, usually with four
    spaces.
```

If you ever need to work with a *DESCRIPTION* file programmatically, take a look at the desc package (*https://desc.r-lib.org*), which usesthis uses heavily under-the-hood.

This chapter shows you how to use the most important DESCRIPTION fields.

---

1 The relationship between "has a *DESCRIPTION* file" and "is a package" is not quite this clear-cut. Many nonpackage projects use a *DESCRIPTION* file to declare their dependencies, i.e., which packages they rely on. In fact, the project for this book does exactly this! This off-label use of *DESCRIPTION* makes it easy to piggy-back on package development tooling to install all the packages necessary to work with a nonpackage project.

Title and Description: What Does Your Package Do?

The `Title` and `Description` fields describe what the package does. They differ only in length:

- `Title` is a one-line description of the package and is often shown in a package listing. It should be plain text (no markup), capitalized like a title, and *not* end in a period. Keep it short: listings will often truncate the title to 65 characters.

- `DESCRIPTION` is more detailed than the title. You can use multiple sentences, but you are limited to one paragraph. If your description spans multiple lines (and it should!), each line must be no more than 80 characters wide. Indent subsequent lines with 4 spaces.

The `Title` and `DESCRIPTION` for ggplot2 are:

```
Title: Create Elegant Data Visualizations Using the Grammar of Graphics
Description: A system for 'declaratively' creating graphics,
    based on "The Grammar of Graphics". You provide the data, tell 'ggplot2'
    how to map variables to aesthetics, what graphical primitives to use,
    and it takes care of the details.
```

A good title and description are important, especially if you plan to release your package to CRAN, because they appear on the package's CRAN landing page as shown in Figure 9-1.

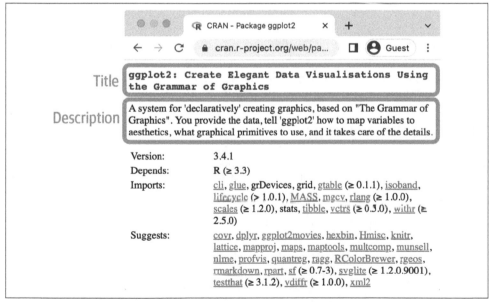

Figure 9-1. How Title and Description appear on ggplot2's CRAN page

Submitting to CRAN

Both the `Title` and `Description` are a frequent source of rejections for reasons not covered by the automated `R CMD check`. In addition to the basics already discussed, here are a few more tips:

- Put the names of R packages, software, and APIs inside single quotes. This goes for both the `Title` and the `Description`. See the ggplot2 example in Figure 9-1.

- If you need to use an acronym, try to do so in `Description`, not in `Title`. In either case, explain the acronym in `Description`, i.e., fully expand it.

- Don't include the package name, especially in `Title`, which is often prefixed with the package name.

- Do not start with "A package for …" or "This package does …". This rule makes sense once you look at the list of CRAN packages by name (*https://oreil.ly/eXcIj*). The information density of such a listing is much higher without a universal prefix like "A package for …".

If these constraints give you writer's block, it often helps to spend a few minutes reading the `Title` and `Description` of packages already on CRAN. Once you read a couple dozen, you can usually find a way to say what you want to say about your package that is also likely to pass CRAN's human-enforced checks.

You'll notice that *DESCRIPTION* gives you only a small amount of space to describe what your package does. This is why it's so important to also include a *README.md* file that goes into much more depth and shows a few examples. You'll learn about that in "README" on page 273.

Author: Who Are You?

Use the `Authors@R` field to identify the package's author, and whom to contact if something goes wrong. This field is unusual because it contains executable R code rather than plain text. Here's an example:

```
Authors@R: person("Hadley", "Wickham", email = "hadley@posit.co",
  role = c("aut", "cre"))

person("Hadley", "Wickham", email = "hadley@posit.co",
  role = c("aut", "cre"))
#> [1] "Hadley Wickham <hadley@posit.co> [aut, cre]"
```

This command says that Hadley Wickham is both the maintainer (`cre`) and an author (`aut`) and that his email address is `hadley@posit.co`. The `person()` function has four main inputs:

- The name, specified by the first two arguments, `given` and `family` (these are normally supplied by position, not name). In English cultures, `given` (first name) comes before `family` (last name). In many cultures, this convention does not hold. For a nonperson entity, such as "R Core Team" or "Posit Software, PBC," use the `given` argument (and omit `family`).

- The `email` address, which is an absolute requirement only for the maintainer. It's important to note that this is the address CRAN uses to let you know if your package needs to be fixed in order to stay on CRAN. Make sure to use an email address that's likely to be around for a while. CRAN policy requires that this be for a person, as opposed to, e.g., a mailing list.

- One or more three-letter codes specifying the `role`. These are the most important roles to know about:

  `cre`
  > The creator or maintainer, the person you should bother if you have problems. Despite being short for "creator," this is the correct role to use for the current maintainer, even if they are not the initial creator of the package.

  `aut`
  > Authors, those who have made significant contributions to the package.

  `ctb`
  > Contributors, those who have made smaller contributions, like patches.

  `cph`
  > Copyright holder. This is used to list additional copyright holders who are not authors, typically companies, like an employer of one or more of the authors.

  `fnd`
  > Funder, the people or organizations that have provided financial support for the development of the package.

- The optional `comment` argument has become more relevant, since `person()` and CRAN landing pages have gained some nice features around ORCID identifiers (*https://orcid.org*). Here's an example of such usage (note the autogenerated URI):

```
person(
  "Jennifer", "Bryan",
  email = "jenny@posit.co",
  role = c("aut", "cre"),
  comment = c(ORCID = "0000-0002-6983-2759")
)
#> [1] "Jennifer Bryan <jenny@posit.co> [aut, cre]
#> (<https://orcid.org/0000-0002-6983-2759>)"
```

You can list multiple authors with `c()`:

```
Authors@R: c(
    person("Hadley", "Wickham", email = "hadley@posit.co", role = "cre"),
    person("Jennifer", "Bryan", email = "jenny@posit.co", role = "aut"),
    person("Posit Software, PBC", role = c("cph", "fnd")))
```

Every package must have at least one author (`aut`) and one maintainer (`cre`), and they might be the same person. The maintainer (`cre`) must have an email address. These fields are used to generate the basic citation for the package (e.g., `citation("pkgname")`). Only people listed as authors will be included in the auto-generated citation (see "Package Citation" on page 117). There are a few extra details if you're including code that other people have written, which you can learn about in "Code You Bundle" on page 178.

An older, still valid approach is to have separate `Maintainer` and `Author` fields in *DESCRIPTION*. However, we strongly recommend the more modern approach of `Authors@R` and the `person()` function, because it offers richer metadata for various downstream uses.

URL and BugReports

As well as the maintainer's email address, it's a good idea to list other places people can learn more about your package. The `URL` field is commonly used to advertise the package's website (see Chapter 19) and to link to a public source repository, where development happens. Multiple URLs are separated with a comma. `BugReports` is the URL where bug reports should be submitted, e.g., as GitHub issues. For example, devtools has:

```
URL: https://devtools.r-lib.org/, https://github.com/r-lib/devtools
BugReports: https://github.com/r-lib/devtools/issues
```

If you use `usethis::use_github()` to connect your local package to a remote GitHub repository, it will automatically populate `URL` and `BugReports` for you. If a package is already connected to a remote GitHub repository, `usethis::use_github_links()` can be called to just add the relevant links to *DESCRIPTION*.

The License Field

The `License` field is mandatory and must specify your package's license in a standard form recognized by R. The official tooling aims to identify standard open source licenses, so it's important to appreciate that `License` is basically a machine-readable field. See Chapter 12 for a full discussion.

Imports, Suggests, and Friends

Two of the most important and commonly used `DESCRIPTION` fields are `Imports` and `Suggests`, which list other packages that your package depends on. Packages listed in `Imports` are needed by your users at runtime and will be installed (or potentially updated) when users install your package via `install.packages()`. The following lines indicate that your package absolutely needs both dplyr and tidyr to work:

```
Imports:
    dplyr,
    tidyr
```

Packages listed in `Suggests` are either needed for development tasks or might unlock optional functionality for your users. The following lines indicate that, while your package can take advantage of ggplot2 and testthat, they're not absolutely required:

```
Suggests:
    ggplot2,
    testthat
```

Both `Imports` and `Suggests` take a comma-separated list of package names. We recommend putting one package on each line, and keeping them in alphabetical order. A non-haphazard order makes it easier for humans to parse this field and appreciate changes.

The easiest way to add a package to `Imports` or `Suggests` is with `usethis::use_package()`. If the dependencies are already in alphabetical order, `use_package()` will keep it that way. In general, it can be nice to run `usethis::use_tidy_description()` regularly, which orders and formats `DESCRIPTION` fields according to a fixed standard.

If you add packages to DESCRIPTION with usethis::use_package(), it will also remind you of the recommended way to call them (explained more in Chapter 11):

```
usethis::use_package("dplyr") # Default is "Imports"
#> ✓ Adding 'dplyr' to Imports field in DESCRIPTION
#> • Refer to functions with `dplyr::fun()`

usethis::use_package("ggplot2", "Suggests")
#> ✓ Adding 'ggplot2' to Suggests field in DESCRIPTION
#> • Use `requireNamespace("ggplot2", quietly = TRUE)` to test if package
#>   is installed
#> • Then directly refer to functions with `ggplot2::fun()`
```

Minimum Versions

If you need a specific version of a package, specify it in parentheses after the package name:

```
Imports:
    dplyr (>= 1.0.0),
    tidyr (>= 1.1.0)
```

The usethis::use_package() convenience function also helps you to set a minimum version:

```
# specific version
usethis::use_package("dplyr", min_version = "1.0.0")

# min version = currently installed version
usethis::use_package("dplyr", min_version = TRUE)
```

You always want to specify a minimum version (dplyr (>= 1.0.0)) rather than an exact version (dplyr (== 1.0.0)). Since R can't have multiple versions of the same package loaded at the same time, specifying an exact dependency dramatically increases the chance of conflicting versions.[2]

Versioning is most important if you will release your package for use by others. Usually people don't have exactly the same versions of packages installed that you do. If someone has an older package that doesn't have a function your package needs, they'll get an unhelpful error message if your package does not advertise the minimum version it needs. However, if you state a minimum version, they'll automatically get an upgrade when they install your package.

---

2 The need to specify the exact versions of packages, rather than minimum versions, comes up more often in the development of nonpackage projects. The renv package (*https://rstudio.github.io/renv*) provides a way to do this, by implementing project-specific environments (package libraries). renv is a reboot of an earlier package called packrat. If you want to freeze the dependencies of a project at exact versions, use renv instead of (or possibly in addition to) a *DESCRIPTION* file.

Think carefully if you declare a minimum version for a dependency. In some sense, the safest thing to do is to require a version greater than or equal to the package's current version. For public work, this is most naturally defined as the current CRAN version of a package; private or personal projects may adopt some other convention. But it's important to appreciate the implications for people who try to install your package: if their local installation doesn't fulfill all of your requirements around versions, installation will force upgrades of these dependencies. This is desirable if your minimum version requirements are genuine, i.e., your package would be broken otherwise. But if your stated requirements have a less solid rationale, this may be unnecessarily conservative and inconvenient.

In the absence of clear, hard requirements, you should set minimum versions (or not) based on your expected user base, the package versions they are likely to have, and a cost-benefit analysis of being too lax versus too conservative. The de facto policy of the tidyverse team is to specify a minimum version when using a known new feature or when someone encounters a version problem in authentic use. This isn't perfect, but we don't currently have the tooling to do better, and it seems to work fairly well in practice.

Depends and LinkingTo

Three other fields allow you to express more specialized dependencies:

Depends

> Prior to the roll-out of namespaces in R 2.14.0 in 2011, Depends was the only way to "depend" on another package. Now, despite the name, you should almost always use Imports, not Depends. You'll learn why, and when you should still use Depends, in "Whether to Import or Depend" on page 152.
>
> The most legitimate current use of Depends is to state a minimum version for R itself, e.g., Depends: R (>= 4.0.0). Again, think carefully if you do this. This raises the same issues as setting a minimum version for a package you depend on, except the stakes are much higher when it comes to R itself. Users can't simply consent to the necessary upgrade, so, if other packages depend on yours, your minimum version requirement for R can cause a cascade of package installation failures.
>
> - The backports package (*https://cran.r-project.org/package=backports*) is useful if you want to use a function like tools::R_user_dir(), which was introduced in 4.0.0 in 2020, while still supporting older R versions.

- The tidyverse packages officially support the current R version, the devel version, and four previous versions.[3] We proactively test this support in the standard build matrix we use for continuous integration.

- Packages with a lower level of use may not need this level of rigor. The main takeaway is: if you state a minimum of R, you should have a reason and you should take reasonable measures to test your claim regularly.

LinkingTo
> If your package uses C or C++ code from another package, you need to list it here.

Enhances
> Packages listed here are "enhanced" by your package. Typically, this means you provide methods for classes defined in another package (a sort of reverse Sug gests). But it's hard to define what that means, so we don't recommend using Enhances.

An R Version Gotcha

Before we leave this topic, we give a concrete example of how easily an R version dependency can creep in and have a broader impact than you might expect. The saveRDS() function writes a single R object as an *.rds* file, an R-specific format. For almost 20 years, *.rds* files used the "version 2" serialization format. "Version 3" became the new default in R 3.6.0 (released April 2019) and cannot be read by R versions prior to 3.5.0 (released April 2018).

Many R packages have at least one *.rds* file lurking within and, if that gets re-generated with a modern R version, by default, the new *.rds* file will have the "version 3" format. When that R package is next built, such as for a CRAN submission, the required R version is automatically bumped to 3.5.0, signaled by this message:

```
NB: this package now depends on R (>= 3.5.0)
  WARNING: Added dependency on R >= 3.5.0 because serialized objects in
  serialize/load version 3 cannot be read in older versions of R.
  File(s) containing such objects:
    'path/to/some_file.rds'
```

Literally, the *DESCRIPTION* file in the bundled package says Depends: R (>= 3.5.0), even if *DESCRIPTION* in the source package says differently.[4]

---

3 See this blog post (*https://www.tidyverse.org/blog/2019/04/r-version-support*) for more.

4 The different package states, such as source versus bundled, are explained in "Package States" on page 33.

When such a package is released on CRAN, the new minimum R version is viral, in the sense that all packages listing the original package in Imports or even Suggests have, to varying degrees, inherited the new dependency on R >= 3.5.0.

The immediate takeaway is to be very deliberate about the version of *.rds* files until R versions prior to 3.5.0 have fallen off the edge of what you intend to support. This particular *.rds* issue won't be with us forever, but similar issues crop up elsewhere, such as in the standards implicit in compiled C or C++ source code. The broader message is that the more reverse dependencies your package has, the more thought you need to give to your package's stated minimum versions, especially for R itself.

Other Fields

A few other DESCRIPTION fields are heavily used and worth knowing about:

- Version is very important as a way of communicating where your package is in its lifecycle and how it is evolving over time. Learn more in Chapter 21.

- LazyData is relevant if your package makes data available to the user. If you specify LazyData: true, the datasets are lazy-loaded, which makes them more immediately available, i.e., users don't have to use data(). The addition of LazyData: true is handled automatically by usethis::use_data(). More detail is given in Chapter 7.

- Encoding describes the character encoding of files throughout your package. Our tooling will set this to Encoding: UTF-8 as this is the most common encoding in use today, and we are not aware of any reasons to use a different value.

- Collate controls the order in which R files are sourced. This only matters if your code has side effects—most commonly because you're using S4. If needed, Collate is typically generated by roxygen2 through use of the @include tag. See ?roxygen2::update_collate for details.

- VignetteBuilder lists any package that your package needs as a vignette engine. Our recommended vignette workflow is described in "Workflow for Writing a Vignette" on page 260, which will list the knitr package in VignetteBuilder.

- SystemRequirements is where you describe dependencies external to R. This is a plain-text field and does not, for example, actually install or check for anything, so you might need to include additional installation details in your README (see "README" on page 273). The most common usage is in the context of a package with compiled code, where SystemRequirements is used to declare the C++ standard, the need for GNU make, or some other external dependency. Examples:

```
SystemRequirements: C++17
SystemRequirements: GNU make
SystemRequirements: TensorFlow (https://www.tensorflow.org/
```

We discourage the explicit use of the `Date` field, as it is extremely easy to forget to update it if you manage `Date` by hand. This field will be populated in the natural course of bundling the package—e.g., when submitting to CRAN—and we recommend that you just let that happen.

There are many other `DESCRIPTION` fields that are used less frequently. A complete list can be found in the "The DESCRIPTION file" section of *Writing R Extensions* (*https://oreil.ly/Yqw62*).

Custom Fields

There is also some flexibility to create your own fields to add additional metadata. In the narrowest sense, the only restriction is that you shouldn't repurpose the official field names used by R. You should also limit yourself to valid English words, so the field names aren't flagged by the spellcheck.

In practice, if you plan to submit to CRAN, we recommend that any custom field name should start with `Config/`. We'll revisit this later when we explain how `Config/Needs/website` is used to record additional packages needed to build a package's website (see "Package Is a Nonstandard Dependency" on page 165).

You might notice that `create_package()` writes two more fields we haven't discussed yet, relating to the use of the roxygen2 package for documentation:

```
Roxygen: list(markdown = TRUE)
RoxygenNote: 7.2.1
```

You will learn more about these in Chapter 16. The use of these specific field names is basically an accident of history and, if it were redone today, they would follow the `Config/*` pattern recommended previously.

## CHAPTER 10

Dependencies: Mindset and Background

You take a dependency when your package uses functionality from another package (or other external tool). In "Imports, Suggests, and Friends" on page 129, we explained how to declare a dependency on another package by listing it in *DESCRIPTION*, usually in `Imports` or `Suggests`. But that still leaves many issues for you to think about:

- When should you take a dependency? What are the risks and rewards? In "When Should You Take a Dependency?" on page 136 we provide a framework for deciding whether a dependency is worth it. This chapter also includes specific sections for deciding between `Imports` and `Suggests` (see "Whether to Import or Suggest" on page 141) and between Imports and `Depends` (see "Whether to Import or Depend" on page 152).

- How should you use different kinds of dependencies in different contexts? That is, imported versus suggested packages, used inside your functions versus tests versus documentation. We have to defer this to the next chapter (Chapter 11), because the justification for those recommendations relies on some additional technical background that we develop here.

A key concept for understanding how packages are meant to work together is that of a namespace (see "Namespace" on page 142). Although it can be a bit confusing, R's namespace system is vital for the package ecosystem. It is what ensures that other packages won't interfere with your code, that your code won't interfere with other packages, and that your package works regardless of the environment in which it's run. We will show how the namespace system works alongside and in concert with the user's search path (see "Search Path" on page 146).

This chapter contains material that could be skipped (or skimmed) when making your first package, when you're probably happy just to make a package that works! But you'll want to revisit the material in this chapter as your packages get more ambitious and sophisticated.

When Should You Take a Dependency?

This section is adapted from the "It Depends" blog post (https://oreil.ly/B8-Wp) and talk (https://oreil.ly/2uQZL) authored by Jim Hester.

Software dependencies are a double-edged sword. On one hand, they let you take advantage of others' work, giving your software new capabilities and making its behavior and interface more consistent with other packages. By using a preexisting solution, you avoid reimplementing functionality, which eliminates many opportunities for you to introduce bugs. On the other hand, your dependencies will likely change over time, which could require you to make changes to your package, potentially increasing your maintenance burden. Your dependencies can also increase the time and disk space needed when users install your package. These downsides have led some to suggest a "dependency zero" mindset. We feel that this is bad advice for most projects and likely to lead to lower functionality, increased maintenance, and new bugs.

Dependencies Are Not Equal

One problem with simply minimizing the absolute number of dependencies is that it treats all dependencies as equivalent, as if they all have the same costs and benefits (or even, infinite cost and no benefit). However, in reality, this is far from the truth. There are many axes upon which dependencies can differ, but some of the most important include:

The type of the dependency
> Some dependencies come bundled with R itself (e.g., base, utils, stats) or are one of the "Recommended" packages (e.g., Matrix, survival). These packages are very low cost to depend on, as they are (nearly) universally installed on all users' systems, and mostly they change only with new R versions. In contrast, there is a higher cost for a dependency that comes from, e.g., a non-CRAN repository, which requires users to configure additional repositories before installation.

The number of upstream dependencies, i.e., recursive dependencies

For example, the rlang package (*https://rlang.r-lib.org*) is intentionally managed as a low-level package and has no upstream dependencies apart from R itself. At the other extreme, there are packages on CRAN with ~250 recursive dependencies.

Already fulfilled dependencies

If your package depends on dplyr, then taking a dependency on tibble does not change the dependency footprint, as dplyr itself already depends on tibble. Additionally, some of the most popular packages (e.g., ggplot2) will already be installed on the majority of users' machines. So adding a ggplot2 dependency is unlikely to incur additional installation costs in most cases.

The burden of installing the package

Various factors make a package more costly to install, in terms of time, space, and human aggravation:

Time to compile

Packages that contain C/C++ can take very different amounts of time to install depending on the complexity of the code. For example, the glue package (*https://glue.tidyverse.org*) takes ~5 seconds to compile on CRAN's build machines, whereas the readr package (*https://readr.tidyverse.org*) takes ~100 seconds to install on the same machines.

Binary package size

Users installing binary packages need to download them, so the size of the binary is relevant, particularly for those with slow internet connections. This also varies a great deal across packages. The smallest packages on CRAN are around 1 Kb in size, while the h2o package (*https://cran.r-project.org/package=h2o*) is 170 Mb, and there are Bioconductor binaries that are over 4 Gb!

System requirements

Some packages require additional system dependencies in order to be used. For instance, the rjags package (*https://cran.r-project.org/package=rjags*) requires a matching installation of the JAGS library. Another example is rJava (*https://cran.r-project.org/package=rJava*), which requires a Java SDK and also has additional steps needed to configure R for the proper Java installation, which has caused installation issues for many people (*https://oreil.ly/T83MM*).

Maintenance capacity

It is reasonable to have higher confidence in a package that is well-established and that is maintained by developers or teams with a long track record and that maintain many other packages. This increases the likelihood that the package will remain on CRAN without interruptions and that the maintainer has an intentional approach to the software lifecycle (see Chapter 21).

Functionality

Some packages implement a critical piece of functionality that is used across many packages. In the tidyverse, broadly defined, the rlang, tidyselect, vctrs, and tibble packages are all examples of this. By using these packages for tricky tasks like nonstandard evaluation or manipulation of vectors and data frames, package authors can avoid reimplementing basic functionality. It's easy to think "how hard can it be to write my own X?" when you are focused on the Happy Path.[1] But a huge part of the value brought by packages like vctrs or tibble is letting someone else worry about edge cases and error handling.[2] There is also value in having shared behavior with other packages, e.g., the tidyverse rules for name repair (*https://oreil.ly/uxzxc*) or recycling (*https://oreil.ly/s0_Ao*).

These specifics hopefully make it clear that package dependencies are not equal.

Prefer a Holistic, Balanced, and Quantitative Approach

Instead of striving for a minimal number of dependencies, we recommend a more holistic, balanced, and quantitative approach.

A holistic approach looks at the project as a whole and asks "who is the primary audience?" If the audience is other package authors, then a leaner package with fewer dependencies may be more appropriate. If, instead, the target user is a data scientist or statistician, they will likely already have many popular dependencies installed and would benefit from a more feature-full package.

A balanced approach understands that adding (or removing) dependencies comes with trade-offs. Adding a dependency gives you additional features, bug fixes, and real-world testing, at the cost of increased installation time, disk space, and maintenance, if the dependency has breaking changes. In some cases it makes sense to *increase* dependencies for a package, even if an implementation already exists. For instance, base R has a number of different implementations of nonstandard evalua-

---

1 In programming, the Happy Path is the scenario where all the inputs make sense and are exactly how things "should be." The Unhappy Path is everything else (objects of length or dimension zero, objects with missing data or dimensions or attributes, objects that don't exist, etc.).

2 Before writing your own version of X, have a good look at the bug tracker and test suite for another package that implements X. This can be useful for appreciating what is actually involved.

tion with varying semantics across its functions. The same used to be true of tidyverse packages as well, but now they all depend on the implementations in the tidyselect (*https://tidyselect.r-lib.org*) and rlang (*https://rlang.r-lib.org*) packages. Users benefit from the improved consistency of this feature and individual package developers can let the maintainers of tidyselect and rlang worry about the technical details.

In contrast, removing a dependency lowers installation time, disk space, and avoids potential breaking changes. However, it means your package will have fewer features or that you must reimplement them yourself. That, in turn, takes development time and introduces new bugs. One advantage of using an existing solution is that you'll get the benefit of all the bugs that have already been discovered and fixed. Especially if the dependency is relied on by many other packages, this is a gift that keeps on giving.

Similar to optimizing performance, if you are worried about the burden of dependencies, it makes sense to address those concerns in a specific and quantitative way. The experimental itdepends package (*https://github.com/r-lib/itdepends*) was created for the talk (*https://oreil.ly/f7W4g*) and blog post (*https://oreil.ly/ESju6*) this section is based on. It is still a useful source of concrete ideas (and code) for analyzing how heavy a dependency is. The pak package (*https://pak.r-lib.org/*) also has several functions that are useful for dependency analysis:

```
pak::pkg_deps_tree("tibble")
#> tibble 3.1.8 ✨
#> ├─fansi 1.0.3 ✨
#> ├─lifecycle 1.0.3 ✨
#> │ ├─cli 3.4.1 ✨ ⬇ (1.28 MB)
#> │ ├─glue 1.6.2 ✨
#> │ └─rlang 1.0.6 ✨ ⬇ (1.81 MB)
#> ├─magrittr 2.0.3 ✨
#> ├─pillar 1.8.1 ✨ ⬇ (673.95 kB)
#> │ ├─cli
#> │ ├─fansi
#> │ ├─glue
#> │ ├─lifecycle
#> │ ├─rlang
#> │ ├─utf8 1.2.2 ✨
#> │ └─vctrs 0.5.1 ✨ ⬇ (1.82 MB)
#> │   ├─cli
#> │   ├─glue
#> │   ├─lifecycle
#> │   └─rlang
#> ├─pkgconfig 2.0.3 ✨
#> ├─rlang
#> └─vctrs
#>
#> Key: ✨ new | ⬇ download

pak::pkg_deps_explain("tibble", "rlang")
```

```
#> tibble -> lifecycle -> rlang
#> tibble -> pillar -> lifecycle -> rlang
#> tibble -> pillar -> rlang
#> tibble -> pillar -> vctrs -> lifecycle -> rlang
#> tibble -> pillar -> vctrs -> rlang
#> tibble -> rlang
#> tibble -> vctrs -> lifecycle -> rlang
#> tibble -> vctrs -> rlang
```

Dependency Thoughts Specific to the tidyverse

The packages maintained by the tidyverse team play different roles in the ecosystem and are managed accordingly. For example, the tidyverse and devtools packages are essentially metapackages that exist for the convenience of an end user. Consequently, it is recommended that other packages *should not depend* on tidyverse[3] or devtools (see "devtools, usethis, and You" on page 27), i.e., these two packages should almost never appear in Imports. Instead, a package maintainer should identify and depend on the specific package that actually implements the desired functionality.

In the previous section, we talked about different ways to gauge the weight of a dependency. Both the tidyverse and devtools can be seen as heavy due to the very high number of recursive dependencies:

```
n_hard_deps <- function(pkg) {
  deps <- tools::package_dependencies(pkg, recursive = TRUE)
  sapply(deps, length)
}

n_hard_deps(c("tidyverse", "devtools"))
#> tidyverse  devtools
#>       114       101
```

In contrast, several packages are specifically conceived as low-level packages that implement features that should work and feel the same across the whole ecosystem. At the time of writing, these include:

- rlang, to support tidy eval and throw errors
- cli and glue, for creating a rich user interface (which includes errors)
- withr, for managing state responsibly
- lifecycle, for managing the lifecycle of functions and arguments

These are basically regarded as free dependencies and can be added to *DESCRIPTION* via usethis::use_tidy_dependencies() (which also does a few more things).

---

3 There is a blog post (*https://oreil.ly/PShpm*) that warns people away from depending on the tidyverse package.

It should come as no surprise that these packages have a very small dependency footprint:

```
tools::package_dependencies(c("rlang", "cli", "glue", "withr", "lifecycle"))
#> $rlang
#> [1] "utils"
#>
#> $cli
#> [1] "utils"
#>
#> $glue
#> [1] "methods"
#>
#> $withr
#> [1] "graphics"  "grDevices" "stats"
#>
#> $lifecycle
#> [1] "cli"   "glue"   "rlang"
```

Submitting to CRAN

Under certain configurations, including those used for incoming CRAN submissions, R CMD check issues a NOTE if there are 20 or more "non-default" packages in Imports:

```
N  checking package dependencies (1.5s)
   Imports includes 29 non-default packages.
   Importing from so many packages makes the package
   vulnerable to any of them becoming unavailable.
   Move as many as possible to Suggests and use
   conditionally.
```

Our best advice is to try hard to comply, as it should be rather rare to need so many dependencies and it's best to eliminate any NOTE that you can. Of course, there are exceptions to every rule and perhaps your package is one them. In that case, you may need to argue your case. It is certainly true that many CRAN packages violate this threshold.

Whether to Import or Suggest

The withr package (*https://withr.r-lib.org*) is a good case study for deciding whether to list a dependency in Imports or Suggests. Withr is very useful for writing tests that clean up after themselves. Such usage is compatible with listing withr in Suggests, since regular users don't need to run the tests. But sometimes a package might also use withr in its own functions, perhaps to offer its own with_*() and local_*() functions. In that case, withr should be listed in Imports.

`Imports` and `Suggests` differ in the strength and nature of dependency:

Imports

Packages listed here *must* be present for your package to work. Any time your package is installed, those packages will also be installed, if not already present. `devtools::load_all()` also checks that all packages in `Imports` are installed.

It's worth pointing out that adding a package to `Imports` ensures it will be installed and *that is all it does*. It has nothing to do with actually importing functions from that package. See "Package Is Listed in Imports" on page 157 for more about how to use a package in `Imports`.

Suggests

Your package can use these packages, but doesn't require them. You might use suggested packages for example datasets to run tests, build vignettes, or maybe there's only one function that needs the package.

Packages listed in `Suggests` are not automatically installed along with your package. This means that you can't assume that your users have installed all the suggested packages, but you can assume that developers have. See "Package Is Listed in Suggests" on page 161 for how to check whether a suggested package is installed.

`Suggests` isn't terribly relevant for packages where the user base is approximately equal to the development team or for packages that are used in a very predictable context. In that case, it's reasonable to just use `Imports` for everything. Using `Suggests` is mostly a courtesy to external users or to accommodate very lean installations. It can free users from downloading rarely needed packages (especially those that are tricky to install) and lets them get started with your package as quickly as possible.

Namespace

So far, we've explained the mechanics of declaring a dependency in DESCRIPTION (see "Imports, Suggests, and Friends" on page 129) and how to analyze the costs and benefits of dependencies (see "When Should You Take a Dependency?" on page 136). Before we explain how to use your dependencies in various parts of your package in Chapter 11, we need to establish the concepts of a package namespace and the search path.

Motivation

As the name suggests, namespaces provide "spaces" for "names." They provide a context for looking up the value of an object associated with a name.

Without knowing it, you've probably already used namespaces. Have you ever used the :: operator? It disambiguates functions with the same name. For example, both the lubridate and here packages provide a here() function. If you attach lubridate, then here, here() will refer to the here version, because the last package attached wins. But if you attach the packages in the opposite order, here() will refer to the lubridate version:

```
library(lubridate)   |   library(here)
library(here)        |   library(lubridate)

here() # here::here() |   here() # lubridate::here()
```

This can be confusing. Instead, you can qualify the function call with a specific namespace: lubridate::here() and here::here(). Then the order in which the packages are attached won't matter:[4]

```
lubridate::here() # always gets lubridate::here()
here::here()      # always gets here::here()
```

As you will see in "Package Is Listed in Imports" on page 157, the package::func tion() calling style is also our default recommendation for how to use your dependencies in the code below *R/*, because it eliminates all ambiguity.

But, in the context of package code, the use of :: is not really our main line of defense against the confusion seen in the previous example. In packages, we rely on namespaces to ensure that every package works the same way regardless of what packages are attached by the user.

Consider the sd() function from the stats package that is part of base R:

```
sd
#> function (x, na.rm = FALSE)
#> sqrt(var(if (is.vector(x) || is.factor(x)) x else as.double(x),
#>     na.rm = na.rm))
#> <bytecode: 0x7fd700fb78b0>
#> <environment: namespace:stats>
```

---

4 We're going to stay focused on packages in this book, but there are other ways than using :: to address conflicts in end-user code: the conflicted package (*https://conflicted.r-lib.org*) and the "conflicts.policy" option (*https://oreil.ly/Fb0HO*) introduced in base R v3.6.0.

It's defined in terms of another function, var(), also from the stats package. So what happens if we override var() with our own definition? Does it break sd()?

```
var <- function(x) -5
var(1:5)
#> [1] -5

sd(1:5)
#> [1] 1.58
```

Surprisingly, it does not! That's because when sd() looks for an object called var(), it looks first in the stats package namespace, so it finds stats::var(), not the var() we created in the global environment. It would be chaos if functions like sd() could be broken by a user redefining var() or by attaching a package that overrides var(). The package namespace system is what saves us from this fate.

The NAMESPACE File

The *NAMESPACE* file plays a key role in defining your package's namespace. Here are selected lines from the *NAMESPACE* file in the testthat package:

```
# Generated by roxygen2: do not edit by hand

S3method(compare,character)
S3method(print,testthat_results)
export(compare)
export(expect_equal)
import(rlang)
importFrom(brio,readLines)
useDynLib(testthat, .registration = TRUE)
```

The first line announces that this file is not written by hand, but rather is generated by the roxygen2 package. We'll return to this topic soon, after we discuss the remaining lines.

You can see that the *NAMESPACE* file looks a bit like R code (but it is not). Each line contains a *directive*: S3method(), export(), importFrom(), and so on. Each directive describes an R object and says whether it's exported from this package to be used by others, or it's imported from another package to be used internally.

These directives are the most important in our development approach, in order of frequency:

export()
    Exports a function (including S3 and S4 generics).

S3method()
    Exports an S3 method.

`importFrom()`
  Imports selected object from another namespace (including S4 generics).

`import()`
  Imports all objects from another package's namespace.

`useDynLib()`
  Registers routines from a DLL (this is specific to packages with compiled code).

There are other directives that we won't cover here, because they are explicitly discouraged or they just rarely come up in our development work:

`exportPattern()`
  Exports all functions that match a pattern. We feel it's safer always to use explicit exports, and we avoid the use of this directive.

`exportClasses()`, `exportMethods()`, `importClassesFrom()`, `importMethodsFrom()`
  Export and import S4 classes and methods. We work in the S4 system only when necessary for compatibility with another package, i.e., we generally don't implement methods or classes that we own with S4. Therefore the S4 coverage in this book is very minimal.

In the devtools workflow, the *NAMESPACE* file is not written by hand! Instead, we prefer to generate *NAMESPACE* with the roxygen2 package, using specific tags located in a roxygen comment above each function's definition in the *R/*.R* files (see "NAMESPACE Workflow" on page 156). We will have much more to say about roxygen comments and the roxygen2 package when we discuss package documentation in Chapter 16. For now, we just lay out the reasons we prefer this method of generating the *NAMESPACE* file:

- Namespace tags are integrated into the source code, so when you read the code it's easier to see what's being exported and imported and why.

- Roxygen2 abstracts away some of the details of *NAMESPACE*. You only need to learn one tag, `@export`, and roxygen2 will figure out which specific directive to use, based on whether the associated object is a regular function, S3 method, S4 method, or S4 class.

- Roxygen2 keeps *NAMESPACE* tidy. No matter how many times `@importFrom foo bar` appears in your roxygen comments, you'll only get one `importFrom(foo, bar)` in your *NAMESPACE*. Roxygen2 also keeps *NAMESPACE* organized in a principled order, sorting first by the directive type and then alphabetically. Roxygen2 takes away the burden of writing *NAMESPACE*, while also trying to keep the file as readable as possible. This organization also makes Git diffs much more informative.

Note that you can choose to use roxygen2 to generate just *NAMESPACE*, just *man/*.Rd* (see Chapter 16), or both (as is our practice). If you don't use any namespace-related tags, roxygen2 won't touch *NAMESPACE*. If you don't use any documentation-related tags, roxygen2 won't touch *man/*.

Search Path

To understand why namespaces are important, you need a solid understanding of search paths. To call a function, R first has to find it. This search unfolds differently for user code than for package code, and that is because of the namespace system.

Function Lookup for User Code

The first place R looks for an object is the global environment. If R doesn't find it there, it looks in the search path, the list of all the packages you have *attached*. You can see this list by running `search()`. For example, here's the search path for the code in this book:

```
search()
#> [1] ".GlobalEnv"        "package:stats"     "package:graphics"
#> [4] "package:grDevices" "package:utils"     "package:datasets"
#> [7] "package:methods"   "Autoloads"         "package:base"
```

This has a specific form (see Figure 10-1):

1. The global environment.

2. The packages that have been attached, e.g., via `library()`, from most recently attached to least.

3. `Autoloads`, a special environment that uses delayed bindings to save memory by only loading package objects (like big datasets) when needed.

4. The base environment, by which we mean the package environment of the base package.

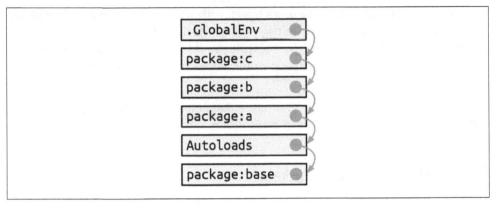

Figure 10-1. Typical state of the search path

Each element in the search path has the next element as its parent, i.e., this is a chain of environments that is searched in order. In the diagram, this relationship is shown as a small blue circle with an arrow that points to the parent. The first environment (the global environment) and the last two (Autoloads and the base environment) are special and maintain their position.

But the middle section of attached packages is more dynamic. When a new package is attached, it is inserted right after and becomes the parent of the global environment. When you attach another package with library(), it changes the search path, as shown in Figure 10-2.

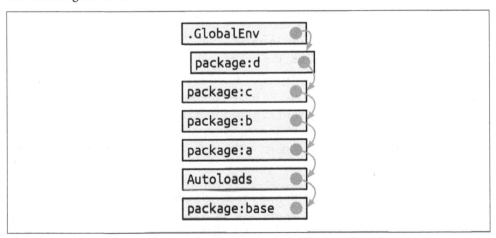

Figure 10-2. A newly attached package is inserted into the search path

The main gotcha around how the user's search path works is the scenario we explored in "Motivation" on page 143, where two packages (lubridate and here) offer competing functions by the same name (here()). It should be very clear now why a user's call to here() can produce a different result, depending on the order in which they attached the two packages.

This sort of confusion would be even more damaging if it applied to package code, but luckily it does not. Now we can explain how the namespace system designs this problem away.

Function Lookup Inside a Package

In "Motivation" on page 143, we proved that a user's definition of a function named var() does *not* break stats::sd(). Somehow, to our immense relief, stats::sd() finds stats::var() when it should. How does that work?

This section is somewhat technical and you can absolutely develop a package with a well-behaved namespace without fully understanding these details. Consider this optional reading that you can consult when and if you're interested. You can learn even more in *Advanced R* (*https://adv-r.hadley.nz*), especially in the chapter on environments, from which we have adapted some of this material.

Every function in a package is associated with a pair of environments—the *package* environment, which is what appears in the user's search path, and the *namespace* environment:

- The package environment is the external interface to the package. It's how a regular R user finds a function in an attached package or with ::. Its parent is determined by search path, i.e., the order in which packages have been attached. The package environment exposes only exported objects.

- The namespace environment is the internal interface of the package. It includes all objects in the package, both exported and nonexported. This ensures that every function can find every other function in the package. Every binding in the package environment also exists in the namespace environment, but not vice versa.

Figure 10-3 depicts the sd() function as a rectangle with a rounded end. The arrows from package:stats and namespace:stats show that sd() is bound in both. But the relationship is not symmetric. The black circle with an arrow pointing back to namespace:stats indicates where sd() will look for objects that it needs: in the namespace environment, not the package environment.

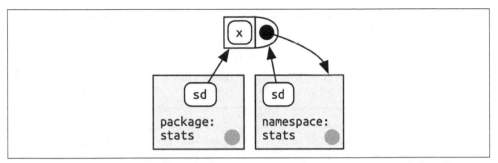

Figure 10-3. An exported function is bound in the package environment and in the namespace, but binds only the namespace

> The package environment controls how users find the function; the namespace controls how the function finds its variables.

Every namespace environment has the same set of ancestors, as depicted in Figure 10-4:

- Each namespace has an *imports* environment that can contain bindings to functions used by the package that are defined in another package. The imports environment is controlled by the package developer with the *NAMESPACE* file. Specifically, directives such as `importFrom()` and `imports()` populate this environment.

- Explicitly importing every base function would be tiresome, so the parent of the imports environment is the base *namespace*. The base namespace contains the same bindings as the base environment, but it has a different parent.

- The parent of the base namespace is the global environment. This means that if a binding isn't defined in the imports environment the package will look for it in the usual way. This is usually a bad idea (because it makes code depend on other loaded packages), so R CMD check automatically warns about such code. It is needed primarily for historical reasons, particularly due to how S3 method dispatch works.

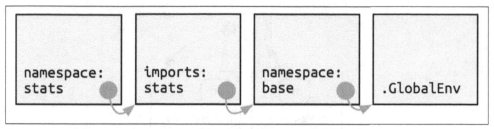

Figure 10-4. The namespace environment has the imports environment as parent, which inherits from the namespace environment of the base package and, ultimately, the global environment

Finally, we can put it all together in this last diagram, Figure 10-5. This shows the user's search path, along the bottom, and the internal stats search path, along the top.

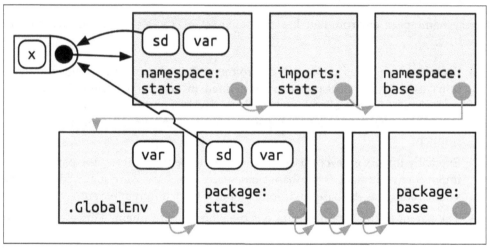

Figure 10-5. For user code, objects are found using the search path, whereas package code uses the namespace

A user (or some package they are using) is free to define a function named var().
But when that user calls sd(), it will always call stats::var() because sd() searches
in a sequence of environments determined by the stats package, not by the user. This
is how the namespace system ensures that package code always works the same way,
regardless of what's been defined in the global environment or what's been attached.

Attaching Versus Loading

It's common to hear something like "we use library(somepackage) to load somepackage." But technically library() *attaches* a package to the search path. This casual
abuse of terminology is often harmless and can even be beneficial in some settings.
But sometimes it's important to be precise and pedantic, and this is one of those
times. Package developers need to know the difference between attaching and loading
a package and when to care about this difference.

If a package is installed:

- *Loading* will load code, data, and any DLLs; register S3 and S4 methods; and run
  the .onLoad() function. After loading, the package is available in memory, but
  because it's not in the search path, you won't be able to access its components
  without using ::. Confusingly, :: will also load a package automatically if it isn't
  already loaded.

- *Attaching* puts the package in the search path (see "Function Lookup for User
  Code" on page 146). You can't attach a package without first loading it, so
  library() (or require()) load then attach the package. This also runs the .onAt
  tach() function.

There are four functions that make a package available, shown in Table 10-1. They
differ based on whether they load or attach, and what happens if the package is not
found (i.e., throws an error or returns FALSE).

Table 10-1. Functions that load or attach a package

	Throws error	Returns FALSE
Load	loadNamespace("x")	requireNamespace("x", quietly = TRUE)
Attach	library(x)	require(x, quietly = TRUE)

Of the four, these two functions are by far the most useful:

- Use library(x) in, e.g., a data analysis script or a vignette. It will throw an error
  if the package is not installed and will terminate the script. You want to attach the
  package to save typing. Never use library() in package code below *R/* or *tests/*.

- Use `requireNamespace("x", quietly = TRUE)` inside a package if you want to specify different behavior depending on whether a suggested package is installed. Note that this also loads the package. We give examples in "In Code Below R/" on page 161.

`loadNamespace()` is somewhat esoteric and is really needed only for internal R code.

`require(pkg)` is almost never a good idea[5] and, we suspect, may come from people projecting certain hopes and dreams onto the function name. Ironically, `require(pkg)` does not actually require success in attaching pkg, and your function or script will soldier on even in the case of failure. This, in turn, often leads to a very puzzling error much later. If you want to "attach or fail," use `library()`. If you want to check whether pkg is available and proceed accordingly, use `require Namespace("pkg", quietly = TRUE)`.

One reasonable use of `require()` is in an example that uses a package your package `Suggests`, which is further discussed in "In Examples and Vignettes" on page 163.

The `.onLoad()` and `.onAttach()` functions mentioned previously are two of several hooks that allow you to run specific code when your package is loaded or attached (or, even, detached or unloaded). Most packages don't need this, but these hooks are useful in certain situations. See "When You Do Need Side Effects" on page 95 for some use cases for `.onLoad()` and `.onAttach()`.

Whether to Import or Depend

We are now in a position to lay out the difference between between `Depends` and `Imports` in the *DESCRIPTION*. Listing a package in either `Depends` or `Imports` ensures that it's installed when needed. The main difference is that a package you list in `Imports` will just be *loaded* when you use it, whereas a package you list in `Depends` will be *attached* when your package is attached.

Unless there is a good reason otherwise, you should always list packages in `Imports`, not `Depends`. That's because a good package is self-contained and minimizes changes to the global landscape, including the search path.[6]

---

5 The classic blog post "`library()` vs `require()` in R" (*https://oreil.ly/kE65s*) by Yihui Xie is another good resource on this.

6 Thomas Leeper created several example packages to demonstrate the puzzling behavior that can arise when packages use `Depends` and shared the work at *https://github.com/leeper/Depends*. This demo also underscores the importance of using `::` (or `importFrom()`) when using external functions in your package, as recommended in Chapter 11.

Users of devtools are actually regularly exposed to the fact that devtools `Depends` on usethis:

```
library(devtools)
#> Loading required package: usethis

search()
#> [1] ".GlobalEnv"        "package:devtools"  "package:usethis"
#> ...
```

This choice is motivated by backward compatibility. When devtools was split into several smaller packages (see "devtools, usethis, and You" on page 27), many of the user-facing functions moved to usethis. Putting usethis in Depends was a pragmatic choice to insulate users from keeping track of which function ended up where.

A more classic example of `Depends` is how the censored package (*https://oreil.ly/ BCZVu*) depends on the parsnip (*https://oreil.ly/UUXO_*) and survival (*https://oreil.ly/ xC-wd*) packages. Parsnip provides a unified interface for fitting models, and censored is an extension package for survival analysis. The censored package is not useful without parsnip and survival, so it makes sense to list them in `Depends`.

## CHAPTER 11

Dependencies: In Practice

This chapter presents the practical details of working with your dependencies inside your package. If you need a refresher on any of the background:

- Chapter 9 covers the *DESCRIPTION* file. Listing a dependency in that file, such as in Imports, is a necessary first step when taking a dependency.
- "When Should You Take a Dependency?" on page 136 provides a decision-making framework for dependencies.
- The technical details of package namespaces, the search path, and attaching versus loading are laid out in "Namespace" on page 142, "Search Path" on page 146, and "Attaching Versus Loading" on page 151.

We're finally ready to talk about how to use different types of dependencies within the different parts of your package:

- In your functions, below *R/*
- In your tests, below *tests/testthat*
- In your examples, in the help topics, below *man/*
- In your vignettes and articles, below *vignettes/*

Confusion About Imports

Let's make this crystal clear:

Listing a package in Imports in DESCRIPTION does not "import" that package.

It is natural to assume that listing a package in `Imports` actually "imports" the package, but this is just an unfortunate choice of name for the `Imports` field. The `Imports` field makes sure that the packages listed there are installed when your package is installed. It does not make those functions available to you, e.g., below *R/*, or to your user.

It is neither automatic nor necessarily advisable that a package listed in `Imports` also appears in *NAMESPACE* via `imports()` or `importFrom()`. It is common for a package to be listed in `Imports` in *DESCRIPTION* but not in *NAMESPACE*. The converse is not true. Every package mentioned in *NAMESPACE* must also be present in the `Imports` or `Depends` fields.

Conventions for This Chapter

Sometimes our examples can feature real functions from real packages. But if we need to talk about a generic package or function, here are the conventions we use:

pkg
    The name of your hypothetical package

aaapkg or bbbpkg
    The name of a hypothetical package your package depends on

`aaa_fun()`
    The name of a function exported by aaapkg

NAMESPACE Workflow

In the following sections, we give practical instructions on how (and when) to import functions from another package into yours and how to export functions from your package. The file that keeps track of all this is the *NAMESPACE* file (more details in "The NAMESPACE File" on page 144).

In the devtools workflow and this book, we generate the *NAMESPACE* file from special comments in the *R/*.R* files. Since the package that ultimately does this work is roxygen2, these are called "roxygen comments." These roxygen comments are also the basis for your package's help topics, which is covered in "The Documentation Workflow" on page 234.

The *NAMESPACE* file starts out with a single commented-out line explaining the situation (and hopefully discouraging any manual edits):

```
# Generated by roxygen2: do not edit by hand
```

As you incorporate roxygen tags to export and import functions, you need to regenerate the *NAMESPACE* file periodically. Here is the general workflow for regenerating *NAMESPACE* (and your documentation):

1. Add namespace-related tags to the roxygen comments in your *R/*.R* files. This is an artificial example, but it gives you the basic idea:

   ```
   #' @importFrom aaapkg aaa_fun
   #' @import bbbpkg
   #' @export
   foo <- function(x, y, z) {
     ...
   }
   ```

2. Run `devtools::document()` (or press Ctrl/Cmd-Shift-D in RStudio) to "document" your package. By default, two things happen:

   - The help topics in the *man/*.Rd* files are updated (covered in Chapter 16).

   - The *NAMESPACE* file is regenerated. In our example, the *NAMESPACE* file would look like:

     ```
     # Generated by roxygen2: do not edit by hand

     export(foo)
     import(bbbpkg)
     importFrom(aaapkg,aaa_fun)
     ```

Roxygen2 is quite smart and will insert the appropriate directive in *NAMESPACE*, i.e., it can usually determine whether to use `export()` or `S3method()`.

RStudio

Press Ctrl/Cmd-Shift-D to generate your package's *NAMESPACE* (and *man/*.Rd* files). This is also available via Document in the Build menu and pane.

Package Is Listed in Imports

Consider a dependency that is listed in `DESCRIPTION` in `Imports`:

```
Imports:
    aaapkg
```

The code inside your package can assume that aaapkg is installed whenever pkg is installed.

In Code Below R/

Our recommended default is to call external functions using the package::
function() syntax:

```
somefunction <- function(...) {
  ...
  x <- aaapkg::aaa_fun(...)
  ...
}
```

Specifically, we recommend that you default to *not* importing anything from aaapkg
into your namespace. This makes it very easy to identify which functions live outside
of your package, which is especially useful when you read your code in the future.
This also eliminates any concerns about name conflicts between aaapkg and your
package.

Of course there are reasons to make exceptions to this rule and to import something
from another package into yours:

An operator
> You can't call an operator from another package via ::, so you must import it.
> Examples: the null-coalescing operator %||% from rlang or the original pipe %>%
> from magrittr.

A function that you use a lot
> If importing a function makes your code much more readable, that's a good
> enough reason to import it. This literally reduces the number of characters
> required to call the external function. This can be especially handy when generat-
> ing user-facing messages, because it makes it more likely that lines in the source
> correspond to lines in the output.

A function that you call in a tight loop
> There is a minor performance penalty associated with ::. It's on the order of
> 100ns, so it will matter only if you call the function millions of times.

A handy function for your interactive workflow is usethis::use_import_from():

```
usethis::use_import_from("glue", "glue_collapse")
```

The preceding call writes this roxygen tag into the source code of your package:

```
#' @importFrom glue glue_collapse
```

Where should this roxygen tag go? There are two reasonable locations:

- As close as possible to the usage of the external function. With this mindset, you
  would place @importFrom in the roxygen comment above the function in your
  package where you use the external function. If this is your style, you'll have to

do it by hand. We have found that this feels natural at first, but starts to break down as you use more external functions in more places.

- In a central location. This approach keeps all @importFrom tags together, in a dedicated section of the package-level documentation file (which can be created with usethis::use_package_doc(); see "Help Topic for the Package" on page 256). This is what use_import_from() implements. So, in *R/pkg-package.R*, you'll end up with something like this:

```
# The following block is used by usethis to automatically manage
# roxygen namespace tags. Modify with care!
## usethis namespace: start
#' @importFrom glue glue_collapse
## usethis namespace: end
NULL
```

Recall that devtools::document() processes your roxygen comments (see "NAME-SPACE Workflow" on page 156), which writes help topics to *man/*.Rd* and, relevant to our current goal, generates the *NAMESPACE* file. If you use use_import_from(), it does this for you and also calls load_all(), making the newly imported function available in your current session.

The roxygen tag in the preceding code causes this directive to appear in the *NAME-SPACE* file:

```
importFrom(glue, glue_collapse)
```

Now you can use the imported function directly in your code:

```
somefunction <- function(...) {
  ...
  x <- glue_collapse(...)
  ...
}
```

Sometimes you make such heavy use of so many functions from another package that you want to import its entire namespace. This should be relatively rare. In the tidyverse, the package we most commonly treat this way is rlang, which functions almost like a base package for us.

Here is the roxygen tag that imports all of rlang. This should appear somewhere in *R/*.R*, such as the dedicated space described previously for collecting all of your namespace import tags:

```
#' @import rlang
```

After calling devtools::document(), this roxygen tag causes this directive to appear in the *NAMESPACE* file:

```
import(rlang)
```

This is the least recommended solution because it can make your code harder to read (you can't tell where a function is coming from), and if you @import many packages, it increases the chance of function name conflicts. Save this for very special situations.

How to not use a package in Imports

Sometimes you have a package listed in Imports, but you don't actually use it inside your package or, at least, R doesn't think you use it. That leads to a NOTE from R CMD check:

```
* checking dependencies in R code ... NOTE
Namespace in Imports field not imported from: 'aaapkg'
  All declared Imports should be used.
```

This can happen if you need to list an indirect dependency in Imports, perhaps to state a minimum version for it. The tidyverse metapackage has this problem on a large scale, since it exists mostly to install a bundle of packages at specific versions. Another scenario is when your package uses a dependency in such a way that requires another package that is only suggested by the direct dependency.[1] There are various situations where it's not obvious that your package truly needs every package listed in Imports, but in fact it does.

How can you get rid of this NOTE?

Our recommendation is to put a namespace-qualified reference (not a call) to an object in aaapkg in some file below *R/*, such as a *.R* file associated with package-wide setup:

```
ignore_unused_imports <- function() {
  aaapkg::aaa_fun
}
```

You don't need to call ignore_unused_imports() anywhere. You shouldn't export it. You don't have to actually exercise aaapkg::aaa_fun(). What's important is to access something in aaapkg's namespace with ::.

Another approach you might be tempted to use is to import aaapkg::aaa_fun() into your package's namespace, probably with the roxygen tag @importFrom aaapkg aaa_fun. This does suppress the NOTE, but it also does more. This causes aaapkg to be loaded whenever your package is loaded. In contrast, if you use the approach we recommend, the aaapkg will be loaded only if your user does something that actually requires it. This rarely matters in practice, but it's always nice to minimize or delay the loading of additional packages.

---

1 For example, if your package needs to call ggplot2::geom_hex(), you might choose to list hexbin in Imports, since ggplot2 only lists it in Suggests.

In Test Code

Refer to external functions in your tests just as you refer to them in the code below *R/*. Usually this means you should use `aaapkg::aaa_fun()`. But if you have imported a particular function, either specifically or as part of an entire namespace, you can just call it directly in your test code.

It's generally a bad idea to use `library(aaapkg)` to attach one of your dependencies somewhere in your tests, because it makes the search path in your tests different from how your package actually works. This is covered in more detail in "Remove Tension Between Interactive and Automated Testing" on page 211.

In Examples and Vignettes

If you use a package that appears in `Imports` in one of your examples or vignettes, you'll need to either attach the package with `library(aaapkg)` or use a `aaapkg::aaa_fun()`-style call. You can assume that aaapkg is available, because that's what `Imports` guarantees. Read more in "Dependencies and Conditional Execution" on page 252 and "Special Considerations for Vignette Code" on page 267.

Package Is Listed in Suggests

Consider a dependency that is listed in *DESCRIPTION* in `Suggests`:

```
Suggests:
    aaapkg
```

You *cannot* assume that every user has installed aaapkg (but you can assume that a developer has). Whether a user has aaapkg will depend on how they installed your package. Most of the functions that are used to install packages support a `dependencies` argument that controls whether to install just the hard dependencies or to take a more expansive approach, which includes suggested packages:

```
install.packages(dependencies =)
remotes::install_github(dependencies =)
pak::pkg_install(dependencies =)
```

Broadly speaking, the default is to not install packages in `Suggests`.

In Code Below R/

Inside a function in your own package, check for the availability of a suggested package with `requireNamespace("aaapkg", quietly = TRUE)`. There are two basic scenarios—the dependency is absolutely required or your package offers some sort of fallback behavior:

```
# the suggested package is required
my_fun <- function(a, b) {
  if (!requireNamespace("aaapkg", quietly = TRUE)) {
    stop(
      "Package \"aaapkg\" must be installed to use this function.",
      call. = FALSE
    )
  }
  # code that includes calls such as aaapkg::aaa_fun()
}

# the suggested package is optional; a fallback method is available
my_fun <- function(a, b) {
  if (requireNamespace("aaapkg", quietly = TRUE)) {
    aaapkg::aaa_fun()
  } else {
    g()
  }
}
```

The rlang package has some useful functions for checking package availability:
`rlang::check_installed()` and `rlang::is_installed()`. Here's how the checks
around a suggested package could look if you use rlang:

```
# the suggested package is required
my_fun <- function(a, b) {
  rlang::check_installed("aaapkg", reason = "to use `aaa_fun()`")
  # code that includes calls such as aaapkg::aaa_fun()
}

# the suggested package is optional; a fallback method is available
my_fun <- function(a, b) {
  if (rlang::is_installed("aaapkg")) {
    aaapkg::aaa_fun()
  } else {
    g()
  }
}
```

These rlang functions have handy features for programming, such as vectorization
over pkg, classed errors with a data payload, and, for `check_installed()`, an offer to
install the needed package in an interactive session.

In Test Code

The tidyverse team generally writes tests as if all suggested packages are available.
That is, we use them unconditionally in the tests.

The motivation for this posture is self-consistency and pragmatism. The key package
needed to run tests is testthat and it appears in Suggests, not in Imports or Depends.

Therefore, if the tests are actually executing, that implies an expansive notion of package dependencies has been applied.

Also, empirically, in every important scenario of running R CMD check, the suggested packages are installed. This is generally true for CRAN, and we ensure that it's true in our own automated checks. However, it's important to note that other package maintainers take a different stance and choose to protect all usage of suggested packages in their tests and vignettes.

Sometimes even we make an exception and guard the use of a suggested package in a test. Here's a test from ggplot2, which uses `testthat::skip_if_not_installed()` to skip execution if the suggested sf package is not available:

```
test_that("basic plot builds without error", {
  skip_if_not_installed("sf")

  nc_tiny_coords <- matrix(
    c(-81.473, -81.741, -81.67, -81.345, -81.266, -81.24, -81.473,
      36.234, 36.392, 36.59, 36.573, 36.437, 36.365, 36.234),
    ncol = 2
  )

  nc <- sf::st_as_sf(
    data_frame(
      NAME = "ashe",
      geometry = sf::st_sfc(sf::st_polygon(list(nc_tiny_coords)), crs = 4326)
    )
  )

  expect_doppelganger("sf-polygons", ggplot(nc) + geom_sf() + coord_sf())
})
```

What might justify the use of `skip_if_not_installed()`? In this case, the sf package can be nontrivial to install, and it is conceivable that a contributor would want to run the remaining tests, even if sf is not available.

Finally, note that `testthat::skip_if_not_installed(pkg, minimum_version = "x.y.z")` can be used to conditionally skip a test based on the version of the other package.

In Examples and Vignettes

Another common place to use a suggested package is in an example, and here we often guard with `require()` or `requireNamespace()`. This example is from `ggplot2::coord_map()`. ggplot2 lists the maps package in Suggests:

```
#' @examples
#' if (require("maps")) {
#'   nz <- map_data("nz")
#'   # Prepare a map of NZ
```

```
#'    nzmap <- ggplot(nz, aes(x = long, y = lat, group = group)) +
#'      geom_polygon(fill = "white", colour = "black")
#'
#'    # Plot it in cartesian coordinates
#'    nzmap
#' }
```

An example is basically the only place where we would use require() inside a package. Read more in "Attaching Versus Loading" on page 151.

Our stance regarding the use of suggested packages in vignettes is similar to that for tests. The key packages needed to build vignettes (rmarkdown and knitr) are listed in Suggests. Therefore, if the vignettes are being built, it's reasonable to assume that all of the suggested packages are available. We typically use suggested packages unconditionally inside vignettes.

But if you choose to use suggested packages conditionally in your vignettes, the knitr chunk option eval is very useful for achieving this. See "Special Considerations for Vignette Code" on page 267 for more.

Package Is Listed in Depends

Consider a dependency that is listed in DESCRIPTION in Depends:

```
Depends:
    aaapkg
```

This situation has a lot in common with a package listed in Imports. The code inside your package can assume that aaapkg is installed on the system. The only difference is that aaapkg will be attached whenever your package is.

In Code Below R/ and in Test Code

Your options are exactly the same as using functions from a package listed in Imports:

- Use the aaapkg::aaa_fun() syntax.
- Import an individual function with the @importFrom aaapkg aaa_fun roxygen tag and call aaa_fun() directly.
- Import the entire aaapkg namespace with the @import aaapkg roxygen tag and call any function directly.

The main difference between this situation and a dependency listed in Imports is that it's much more common to import the entire namespace of a package listed in Depends. This often makes sense, due to the special dependency relationship that motivated listing it in Depends in the first place.

In Examples and Vignettes

This is the most obvious difference with a dependency in `Depends` versus `Imports`. Since your package is attached when your examples are executed, so is the package listed in `Depends`. You don't have to attach it explicitly with `library(aaapkg)`.

The ggforce package `Depends` on ggplot2 and the examples for `ggforce::geom_mark_rect()` use functions like `ggplot2::ggplot()` and `ggplot2::geom_point()` without any explicit call to `library(ggplot2)`:

```
ggplot(iris, aes(Petal.Length, Petal.Width)) +
  geom_mark_rect(aes(fill = Species, filter = Species != 'versicolor')) +
  geom_point()
# example code continues ...
```

The first line of code executed in one of your vignettes is probably `library(pkg)`, which attaches your package and, as a side effect, attaches any dependency listed in `Depends`. You do not need to explicitly attach the dependency before using it. The censored package `Depends` on the survival package and the code in `vignette("examples", package = "censored")` starts out like so:

```
library(tidymodels)
library(censored)
#> Loading required package: survival

# vignette code continues ...
```

Package Is a Nonstandard Dependency

In packages developed with devtools, you may see *DESCRIPTION* files that use a couple other nonstandard fields for package dependencies specific to development tasks.

Depending on the Development Version of a Package

The `Remotes` field can be used when you need to install a dependency from a nonstandard place, i.e., from somewhere besides CRAN or Bioconductor. One common example of this is when you're developing against a development version of one of your dependencies. During this time, you'll want to install the dependency from its development repository, which is often GitHub. The way to specify various remote sources is described in a devtools vignette (*https://oreil.ly/tY-Bc*) and in a pak help topic (*https://oreil.ly/5Z40P*).

The dependency and any minimum version requirement still need to be declared in the normal way in, e.g., `Imports`. `usethis::use_dev_package()` helps to make the necessary changes in `DESCRIPTION`. If your package temporarily relies on a development version of aaapkg, the affected `DESCRIPTION` fields might evolve like this:

```
Stable -->                  Dev -->                       Stable again
---------------------       -----------------------       ---------------------
Package: pkg                Package: pkg                  Package: pkg
Version: 1.0.0              Version: 1.0.0.9000           Version: 1.1.0
Imports:                    Imports:                      Imports:
    aaapkg (>= 2.1.3)           aaapkg (>= 2.1.3.9000)        aaapkg (>= 2.2.0)
                            Remotes:
                                jane/aaapkg
```

CRAN

It's important to note that you should not submit your package to CRAN in the intermediate state, meaning with a Remotes field and with a dependency required at a version that's not available from CRAN or Bioconductor. For CRAN packages, this can only be a temporary development state, eventually resolved when the dependency updates on CRAN and you can bump your minimum version accordingly.

Config/Needs/* Field

You may also see devtools-developed packages with packages listed in DESCRIPTION fields in the form of Config/Needs/*, which we described in "Custom Fields" on page 134.

The use of Config/Needs/* is not directly related to devtools. It's more accurate to say that it's associated with continuous integration workflows made available to the community at *https://github.com/r-lib/actions/* and exposed via functions such as usethis::use_github_actions(). A Config/Needs/* field tells the setup-r-dependencies GitHub Action (*https://oreil.ly/w9KLZ*) about extra packages that need to be installed.

Config/Needs/website is the most common, and it provides a place to specify packages that aren't a formal dependency but that must be present in order to build the package's website (see Chapter 19). The readxl package is a good example. It has a nonvignette article on workflows (*https://oreil.ly/K8ZR-*) that shows readxl working in concert with other tidyverse packages, such as readr and purrr. But it doesn't make sense for readxl to have a formal dependency on readr or purrr or (even worse) the tidyverse!

On the left is what readxl has in the Config/Needs/website field of DESCRIPTION to indicate that the tidyverse is needed to build the website, which is also formatted with styling that lives in the tidyverse/template GitHub repo. On the right is the corresponding excerpt from the configuration of the workflow that builds and deploys the website:

```
in DESCRIPTION                          in .github/workflows/pkgdown.yaml
-------------------------               ---------------------------------
Config/Needs/website:                   - uses: r-lib/actions/setup-r-dependencies@v2
    tidyverse,                            with:
    tidyverse/tidytemplate                  extra-packages: pkgdown
                                            needs: website
```

Package websites and continuous integration are discussed more in Chapter 19 and "Continuous Integration" on page 298, respectively.

The `Config/Needs/*` convention is handy because it allows a developer to use *DESCRIPTION* as their definitive record of package dependencies, while maintaining a clean distinction between true runtime dependencies versus those that are needed only for specialized development tasks.

Exports

For a function to be usable outside of your package, you must *export* it. When you create a new package with `usethis::create_package()`, nothing is exported at first, even once you add some functions. You can still experiment interactively with `load_all()`, since that loads all functions, not just those that are exported. But if you install and attach the package with `library(pkg)` in a fresh R session, you'll notice that no functions are available.

What to Export

Export functions that you want other people to use. Exported functions must be documented, and you must be cautious when changing their interface—other people are using them! Generally, it's better to export too little than too much. It's easy to start exporting something that you previously did not; it's hard to stop exporting a function because it might break existing code. Always err on the side of caution and simplicity. It's easier to give people more functionality than it is to take away stuff they're used to.

We believe that packages that have a wide audience should strive to do one thing and do it well. All functions in a package should be related to a single problem (or a set of closely related problems). Any functions not related to that purpose should not be exported. For example, most of our packages have a *utils.R* file ("Organize Functions Into Files" on page 83) that contains small utility functions that are useful internally but aren't part of the core purpose of those packages. We don't export such functions. There are at least two reasons for this:

- Freedom to be less robust and less general. A utility for internal use doesn't have to be implemented in the same way as a function used by others. You just need to cover your own use case.

- Regrettable reverse dependencies. You don't want people depending on your package for functionality and functions that are unrelated to its core purpose.

That said, if you're creating a package for yourself, it's far less important to be this disciplined. Because you know what's in your package, it's fine to have a local "miscellany" package that contains a hodgepodge of functions that you find useful. But it is probably not a good idea to release such a package for wider use.

Sometimes your package has a function that could be of interest to other developers extending your package, but not to typical users. In this case, you want to export the function but also give it a very low profile in terms of public documentation. This can be achieved by combining the roxygen tags @export and @keywords internal. The internal keyword keeps the function from appearing in the package index, but the associated help topic still exists and the function still appears among those exported in the *NAMESPACE* file.

Re-exporting

Sometimes you want to make something available to users of your package that is actually provided by one of your dependencies. When devtools was split into several smaller packages (see "devtools, usethis, and You" on page 27), many of the user-facing functions moved elsewhere. For usethis, the chosen solution was to list it in Depends (see "Whether to Import or Depend" on page 152), but that is not a good general solution. Instead, devtools now re-exports certain functions that actually live in a different package.

Here is a blueprint for re-exporting an object from another package, using the session_info() function as our example:

1. List the package that hosts the re-exported object in Imports in *DESCRIPTION*.[2] In this case, the session_info() function is exported by the sessioninfo package:

    ```
    Imports:
        sessioninfo
    ```

2. In one of your *R/*.R* files, have a reference to the target function, preceded by roxygen tags for both importing and exporting:

    ```
    #' @export
    #' @importFrom sessioninfo session_info
    sessioninfo::session_info
    ```

That's it! Next time you regenerate *NAMESPACE*, these two lines will be there (typically interspersed with other exports and imports):

---

2 Remember usethis::use_package() is helpful for adding dependencies to DESCRIPTION.

```
...
export(session_info)
...
importFrom(sessioninfo,session_info)
...
```

And this explains how `library(devtools)` makes `session_info()` available in the current session. This will also lead to the creation of the *man/reexports.Rd* file, which finesses the requirement that your package must document all of its exported functions. This help topic lists all re-exported objects and links to their primary documentation.

Imports and Exports Related to S3

R has multiple object-oriented programming (OOP) systems:

- S3 is currently the most important for us and is what's addressed in this book. The S3 chapter of *Advanced R* (*https://oreil.ly/HtLIb*) is a good place to learn more about S3 conceptually, and the vctrs package (*https://oreil.ly/2ZlPL*) is worth studying for practical knowledge.

- S4 is very important within certain R communities, most notably within the Bioconductor project. We use S4 only when it's necessary for compatibility with other packages. If you want to learn more, the S4 chapter of *Advanced R* (*https://oreil.ly/4_be0*) is a good starting point and has recommendations for additional resources.

- R6 is used in many tidyverse packages (broadly defined) but is out of scope for this book. Good places to learn more include the R6 package website (*https://oreil.ly/FSawp*), the R6 chapter of *Advanced R* (*https://oreil.ly/A5VdR*), and the roxygen2 documentation related to R6 (*https://oreil.ly/XoNI8*).

In terms of namespace issues around S3 classes, the main things to consider are generic functions and their class-specific implementations known as methods. If your package "owns" an S3 class, it makes sense to export a user-friendly constructor function. This is often just a regular function and there is no special S3 angle.

If your package "owns" an S3 generic and you want others to be able to use it, you should export the generic. For example, the dplyr package exports the generic function `dplyr::count()` and also implements and exports a specific method, `count.data.frame()`:

```
#' ... all the usual documentation for count() ...
#' @export
count <- function(x, ..., wt = NULL, sort = FALSE, name = NULL) {
  UseMethod("count")
}
```

```
#' @export
count.data.frame <- function(x,
                                   ...,
                                   wt = NULL,
                                   sort = FALSE,
                                   name = NULL,
                                   .drop = group_by_drop_default(x)) { ... }
```

The corresponding lines in dplyr's *NAMESPACE* file look like this:

```
...
S3method(count,data.frame)
...
export(count)
...
```

Now imagine that your package implements a method for count() for a class you "own" (not data.frame). A good example is the dbplyr package, which implements count() for the tbl_lazy class. An add-on package that implements an S3 generic for a new class should list the generic-providing package in Imports, import the generic into its namespace, and export its S3 method. Here's part of dbplyr's *DESCRIPTION* file:

```
Imports:
    ...,
    dplyr,
    ...
```

In *dbplyr/R/verb-count.R*, we have:

```
#' @importFrom dplyr count
#' @export
count.tbl_lazy <- function(x, ..., wt = NULL, sort = FALSE, name = NULL) { ... }
```

In *NAMESPACE*, we have:

```
S3method(count,tbl_lazy)
...
importFrom(dplyr,count)
```

dbplyr also provides methods for various generics provided by the base package, such as dim() and names(). In this case, there is no need to import those generics, but it's still necessary to export the methods. In *dbplyr/R/tbl_lazy.R*, we have:

```
#' @export
dim.tbl_lazy <- function(x) {
  c(NA, length(op_vars(x$lazy_query)))
}

#' @export
names.tbl_lazy <- function(x) {
  colnames(x)
}
```

In *NAMESPACE*, this produces:

```
S3method(dim,tbl_lazy)
...
S3method(names,tbl_lazy)
```

The last and trickiest case is when your package offers a method for a generic "owned" by a package you've listed in Suggests. The basic idea is that you want to register the availability of your S3 method conditionally, when your package is being loaded. If the suggested package is present, your S3 method should be registered, but otherwise it should not.

We'll illustrate this with an example. Within the tidyverse, the glue package is managed as a low-level package that should have minimal dependencies (see "Dependency Thoughts Specific to the tidyverse" on page 140). Glue functions generally return a character vector that also has the "glue" S3 class:

```
library(glue)
name <- "Betty"
(ret <- glue('My name is {name}.'))
#> My name is Betty.
class(ret)
#> [1] "glue"      "character"
```

The motivation for this is that it allows glue to offer special methods for print(), the + operator, and subsetting via [and [[. One downside, though, is that this class attribute complicates string comparisons:

```
identical(ret, "My name is Betty.")
#> [1] FALSE
all.equal(ret, "My name is Betty.")
#> [1] "Attributes: < Modes: list, NULL >"
#> [2] "Attributes: < Lengths: 1, 0 >"
#> [3] "Attributes: < names for target but not for current >"
#> [4] "Attributes: < current is not list-like >"
#> [5] "target is glue, current is character"
```

Therefore, for testing, it is helpful if glue offers a method for testthat::compare(), which explains why this expectation succeeds:

```
testthat::expect_equal(ret, "My name is Betty.")
```

But glue can't list testthat in Imports! It must go in Suggests. The solution is to register the method conditionally when glue is loaded. Here is a redacted version of glue's .onLoad() function, where you'll see that it conditionally registers some other methods as well:

```
.onLoad <- function(...) {
  s3_register("testthat::compare", "glue")
  s3_register("waldo::compare_proxy", "glue")
  s3_register("vctrs::vec_ptype2", "glue.glue")
  ...
  invisible()
}
```

The `s3_register()` function comes from the vctrs package. If you don't have an organic need to depend on vctrs, it is common (and encouraged) to simply inline the `s3_register()` source (*https://oreil.ly/X0aux*) into your own package. You can't always copy code from other people's packages and paste it into yours, but you can in this case. This usage is specifically allowed by the license of the source code of `s3_register()`. This provides a great segue into Chapter 12, which is all about licensing.

Licensing

The goal of this chapter is to give you the basic tools to manage licensing for your R package. Obviously, we are R developers and not lawyers, and none of this is legal advice. But fortunately, if you're writing either an open source package or a package used only within your organization,[1] you don't need to be an expert to do the right thing. You need to pick a license that declares how you want your code to be used, and if you include code written by someone else, you need to respect the license that it uses.

This chapter begins with an overview of licensing, and how to license your own code. We'll then discuss the most important details of accepting code given to you (e.g., in a pull request) and how to bundle code written by other people. We'll finish off with a brief discussion of the implications of using code from other packages.

Big Picture

To understand the author's wishes, it's useful to understand the two major camps of open source licenses:

- *Permissive* licenses are very easygoing. Code with a permissive license can be freely copied, modified, and published, and the only restriction is that the license must be preserved. The *MIT* (*https://choosealicense.com/licenses/mit*) and *Apache* (*https://choosealicense.com/licenses/apache-2.0*) licenses are the most common modern permissive licenses; older permissive licenses include the various forms of the *BSD* license.

---

1 If you're selling your package, however, we'd highly recommend that you consult a lawyer.

- *Copyleft* licenses are stricter. The most common copyleft license is the *GPL* (*https://oreil.ly/vHmcP*), which allows you to freely copy and modify the code for personal use, but if you publish modified versions or bundle with other code, the modified version or complete bundle must also be licensed with the GPL.

When you look across all programming languages, permissive licenses are the most common. For example, a 2015 survey of GitHub repositories (*https://oreil.ly/QDSux*) found that ~55% used a permissive license and ~20% used a copyleft license. The R community is rather different: as of 2022, our analysis[2] found that ~70% of CRAN packages use a copyleft license and ~20% use a permissive license. This means licensing your R package requires a little more care than for other languages.

Code You Write

We'll start by talking about code that you write, and how to license it to make clear how you want people to treat it. It's important to use a license because if you don't, the default copyright laws apply, which means that no one is allowed to make a copy of your code without your express permission.

In brief:

- If you want a permissive license so people can use your code with minimal restrictions, choose the MIT license (*https://choosealicense.com/licenses/mit*) with `use_mit_license()`.

- If you want a copyleft license so that all derivatives and bundles of your code are also open source, choose the GPLv3 license (*https://choosealicense.com/licenses/gpl-3.0*) with `use_gpl_license()`.

- If your package primarily contains data, not code, and you want minimal restrictions, choose the CC0 license (*https://choosealicense.com/licenses/cc0-1.0*) with `use_cc0_license()`. Or if you want to require attribution when your data is used, choose the CC BY license (*https://choosealicense.com/licenses/cc-by-4.0*) by calling `use_ccby_license()`.

- If you don't want to make your code open source, call `use_propriet ary_license()`. Such packages cannot be distributed by CRAN.

We'll come back to more details and present a few other licenses in "More Licenses for Code" on page 175.

---

2 Inspired by that of Sean Kross (*https://oreil.ly/6IYBa*).

Key Files

There are three key files used to record your licensing decision:

- Every license sets the `License` field in the *DESCRIPTION*. This contains the name of the license in a standard form so that R CMD check and CRAN can automatically verify it. It comes in four main forms:

  — A name and version specification, e.g., `GPL (>= 2)`, or `Apache License (== 2.0)`.

  — A standard abbreviation, e.g., `GPL-2`, `LGPL-2.1`, `Artistic-2.0`.

  — A name of a license "template" and a file containing specific variables. The most common case is `MIT + file LICENSE`, where the *LICENSE* file needs to contain two fields: the year and copyright holder.

  — Pointer to the full text of a nonstandard license, `file LICENSE`.

    More complicated licensing structures are possible but outside the scope of this text. See the "Licensing" section (*https://oreil.ly/Niq0u*) of *Writing R Extensions* for details.

- As described, the *LICENSE* file is used in one of two ways. Some licenses are templates that require additional details to be complete in the *LICENSE* file. The *LICENSE* file can also contain the full text of nonstandard and non-open source licenses. You are not permitted to include the full text of standard licenses.

- *LICENSE.md* includes a copy of the full text of the license. All open source licenses require a copy of the license to be included, but CRAN does not permit you to include a copy of standard licenses in your package, so we also use *.Rbuildignore* to make sure this file is not sent to CRAN.

There is one other file that we'll come back to in "How to Include" on page 179: *LICENSE.note*. This is used when you have bundled code written by other people, and parts of your package have more permissive licenses than the whole.

More Licenses for Code

We gave you the absolute minimum you need to know. But it's worth mentioning a few more important licenses roughly ordered from most permissive to least permissive:

`use_apache_license()`
  The Apache License (*https://choosealicense.com/licenses/apache-2.0*) is similar to the MIT license, but it also includes an explicit patent grant. Patents are another component of intellectual property distinct from copyrights, and some organizations also care about protection from patent claims.

`use_lgpl_license()`

> The LGPL (*https://choosealicense.com/licenses/lgpl-3.0*) is a little weaker than the GPL, allowing you to bundle LPGL code using any license for the larger work.

`use_gpl_license()`

> We've discussed the GPL (*https://choosealicense.com/licenses/gpl-3.0*) already, but there's one important wrinkle to note: the GPL has two major versions, GPLv2 and GPLv3, and they're not compatible (i.e., you can't bundle GPLv2 and GPLv3 code in the same project). To avoid this problem it's generally recommended to license your package as GPL >=2 or GPL >= 3 so that future versions of the GPL license also apply to your code. This is what `use_gpl_license()` does by default.

`use_agpl_license()`

> The AGPL (*https://choosealicense.com/licenses/agpl-3.0*) defines distribution to include providing a service over a network, so that if you use AGPL code to provide a web service, all bundled code must also be open sourced. Because this is a considerably broader claim than the GPL, many companies expressly forbid the use of AGPL software.

Many other licenses are available. To get a high-level view of the open source licensing space, and the details of individual licenses, we highly recommend *https://choosea license.com*, which we've used in the preceding links. For more details about licensing R packages, we recommend *Licensing R* (*https://thinkr-open.github.io/licensing-r*) by Colin Fay. The primary downside of choosing a license not in the preceding list is that fewer R users will understand what it means, making it harder for them to use your code.

Licenses for Data

All these licenses are designed specifically to apply to source code, so if you're releasing a package that primarily contains data, you should use a different type of license. We recommend one of two Creative Commons (*http://creativecommons.org*) licenses:

- If you want to make the data as freely available as possible, you use the CC0 license with `use_cc0_license()`. This is a permissive license that's equivalent to the MIT license, but it applies to data, not code.[3]

- If you want to require attribution when someone else uses your data, you can use the CC-BY license, with `use_ccby_license()`.

---

[3] If you are concerned about the implications of the CC0 license with respect to citation, you might be interested in the Dryad blog post "Why does Dryad use CC0?" (*https://oreil.ly/61dS7*).

Relicensing

Changing your license after the fact is hard because it requires the permission of all copyright holders, and unless you have taken special steps (more on that shortly) this will include everyone who has contributed a nontrivial amount of code.

If you do need to relicense a package, we recommend the following steps:

1. Check the `Authors@R` field in the *DESCRIPTION* to confirm that the package doesn't contain bundled code (which we'll talk about in "Code You Bundle" on page 178).

2. Find all contributors by looking at the Git history or the contributors display on GitHub.

3. Optionally, inspect the specific contributions and remove people who only contributed typo fixes and similar.[4]

4. Ask every contributor if they're OK with changing the license. If every contributor is on GitHub, the easiest way to do this is to create an issue where you list all contributors and ask them to confirm that they're OK with the change.

5. Once all copyright holders have approved, make the change by calling the appropriate license function.

You can read about how the tidyverse followed this process to unify on the MIT license at *https://oreil.ly/p10Vv*.

Code Given to You

Many packages include code not written by the author. There are two main ways this happens: other people might choose to contribute to your package using a pull request or similar, or you might find some code and choose to bundle it. This section will discuss code that others give to you, and the next section will discuss code that you bundle.

When someone contributes code to your package using a pull request or similar, you can assume that the author is happy for their code to use your license. This is explicit in the GitHub terms of service (*https://oreil.ly/qb2St*) but is generally considered to be true regardless of how the code is contributed.[5]

---

4 Very simple contributions like typo fixes are generally not protected by copyright because they're not creative works. But even a single sentence can be considered a creative work, so err on the side of safety, and if you have any doubts leave the contributor in.

5 Some particularly risk-averse organizations require contributors to provide a developer certificate of origin (*https://developercertificate.org*), but this is relatively rare in general, and we haven't seen it in the R community.

However, the author retains copyright of their code, which means that you can't change the license without their permission. If you want to retain the ability to change the license, you need an explicit "contributor license agreement" or CLA, where the author explicitly reassigns the copyright. This is most important for dual open source/commercial projects because it easily allows for dual licensing where the code is made available to the world with a copyleft license, and to paying customers with a different, more permissive, license.

It's also important to acknowledge the contribution, and it's good practice to be generous with thanks and attribution. In the tidyverse, we ask that all code contributors include a bullet in *NEWS.md* with their GitHub username, and we thank all contributors in release announcements. We add only core developers[6] to the *DESCRIPTION* file; but some projects choose to add all contributors no matter how small the contribution.

Code You Bundle

There are three common reasons that you might choose to bundle code written by someone else:

- You're including someone else's CSS or JS library to create a useful and attractive web page or HTML widgets. Shiny is a great example of a package that does this extensively.

- You're providing an R wrapper for a simple C or C++ library. (For complex C/C++ libraries, you don't usually bundle the code in your package, but instead link to a copy installed elsewhere on the system).

- You've copied a small amount of R code from another package to avoid taking a dependency. Generally, taking a dependency on another package is the right thing to do because you don't need to worry about licensing, and you'll automatically get bug fixes. But sometimes you need only a very small amount of code from a big package, and copying and pasting it into your package is the right thing to do.

License Compatibility

Before you bundle someone else's code into your package, you need to first check that the bundled license is compatible with your license. When distributing code, you can add additional restrictions, but you cannot remove restrictions, which means that

---

6 That is, people responsible for ongoing development. This is best made explicit in the ggplot2 governance document, *GOVERNANCE.md* (*https://oreil.ly/hNtaJ*).

license compatibility is not symmetric. For example, you can bundle MIT licensed code in a GPL licensed package, but you can not bundle GPL licensed code in an MIT licensed package.

There are five main cases to consider:

- If your license and their license are the same: it's OK to bundle.
- If their license is MIT or BSD, it's OK to bundle.
- If their code has a copyleft license and your code has a permissive license, you can't bundle their code. You'll need to consider an alternative approach, either looking for code with a more permissive license or putting the external code in a separate package.
- If the code comes from Stack Overflow, it's licensed (*https://oreil.ly/Iz2Yr*) with the Creative Common CC BY-SA license, which is compatible only with GPLv3 (*https://oreil.ly/Rvsvt*). This means that you need to take extra care when using Stack Overflow code in open source packages. Learn more at *https://oreil.ly/qlyFh*.
- Otherwise, you'll need to do a little research. Wikipedia has a useful diagram (*https://oreil.ly/kv5GQ*), and Google is your friend. It's important to note that different versions of the same license are not necessarily compatible, e.g., GPLv2 and GPLv3 are not compatible.

If your package isn't open source, things are more complicated. Permissive licenses are still easy, and copyleft licenses generally don't restrict use as long as you don't distribute the package outside your company. But this is a complex issue and opinions differ, and you should check with your legal department first.

How to Include

Once you've determined that the licenses are compatible, you can bring the code in your package. When doing so, you need to preserve all existing license and copyright statements and make it as easy as possible for future readers to understand the licensing situation:

- If you're including a fragment of another project, it's generally best to put in its own file and ensure that file has copyright statements and license description at the top.
- If you're including multiple files, put in a directory, and put a *LICENSE* file in that directory.

You also need to include some standard metadata in Authors@R. You should use role = "cph" to declare that the author is a copyright holder, with a comment describing what they're the author of.

If you're submitting to CRAN and the bundled code has a different (but compatible) license, you also need to include a *LICENSE.note* file that describes the overall license of the package, and the specific licenses of each individual component. For example, the diffviewer package bundles six JavaScript libraries all of which use a permissive license. The *DESCRIPTION* (*https://oreil.ly/nkoTK*) lists all copyright holders, and the *LICENSE.note* (*https://oreil.ly/_6Wth*) describes their licenses. (Other packages use other techniques, but we think this is the simplest approach that will fly with CRAN.)

Code You Use

Obviously all the R code you write uses R, and R is licensed with the GPL. Does that mean your R code must always be GPL licensed? No, and the R Foundation made this clear (*https://oreil.ly/PPvyK*) in 2009. Similarly, it's our personal opinion that the license of your package doesn't need to be compatible with the licenses of R packages that you use merely by calling their exported R functions (i.e., via Suggests or Imports).

Things are different in other languages, like C, because creating a C executable almost invariably ends up copying some component of the code you use into the executable. This can also come up if your R package has compiled code and you link to (using the LinkingTo in your *DESCRIPTION*): you'll need to do more investigation to make sure your license is compatible. However, if you're just linking to R itself, you are generally free to license as you wish because R headers are licensed with the Lesser GPL (*https://oreil.ly/7qmHn*).

Of course, any user of your package will have to download all the packages that your package depends on (as well as R itself), so they will still have to comply with the terms of those licenses.

## PART IV

Testing

## CHAPTER 13

Testing Basics

Testing is a vital part of package development: it ensures that your code does what you want. Testing, however, adds an additional step to your workflow. To make this task easier and more effective this chapter will show you how to do formal automated testing using the testthat package.

The first stage of your testing journey is to become convinced that testing has enough benefits to justify the work. For some of us, this is easy to accept. Others must learn the hard way.

Once you've decided to embrace automated testing, it's time to learn some mechanics and figure out where testing fits into your development workflow.

As you and your R packages evolve, you'll start to encounter testing situations where it's fruitful to use techniques that are somewhat specific to testing and differ from what we do below *R/*.

Why Is Formal Testing Worth the Trouble?

Up until now, your workflow probably looks like this:

1. Write a function.
2. Load it with `devtools::load_all()`, maybe via Ctrl/Cmd-Shift-L.
3. Experiment with it in the console to see if it works.
4. Rinse and repeat.

While you *are* testing your code in this workflow, you're only doing it informally. The problem with this approach is that when you come back to this code in 3 months to add a new feature, you've probably forgotten some of the informal tests you ran the first time around. This makes it very easy to break code that used to work.

Many of us embrace automated testing when we realize we're refixing a bug for the second or fifth time. While writing code or fixing bugs, we might perform some interactive tests to make sure the code we're working on does what we want. But it's easy to forget all the different use cases you need to check, if you don't have a system for storing and rerunning the tests. This is a common practice among R programmers. The problem is not that you don't test your code, it's that you don't automate your tests.

In this chapter you'll learn how to transition from informal ad hoc testing, done interactively in the console, to automated testing (also known as unit testing). While turning casual interactive tests into formal tests requires a little more work up front, it pays off in four ways:

Fewer bugs

Because you're explicit about how your code should behave, you will have fewer bugs. The reason is a bit like why double entry book keeping works: because you describe the behavior of your code in two places, both in your code and in your tests, you are able to check one against the other.

With informal testing, it's tempting to just explore typical and authentic usage, similar to writing examples. However, when writing formal tests, it's natural to adopt a more adversarial mindset and to anticipate how unexpected inputs could break your code.

If you always introduce new tests when you add a new feature or function, you'll prevent many bugs from being created in the first place, because you will proactively address pesky edge cases. Tests also keep you from (re-)breaking one feature when you're tinkering with another.

Better code structure

Code that is well designed tends to be easy to test, and you can turn this to your advantage. If you are struggling to write tests, consider if the problem is actually the design of your function(s). The process of writing tests is a great way to get free, private, and personalized feedback on how well-factored your code is. If you integrate testing into your development workflow (versus planning to slap tests on "later"), you'll subject yourself to constant pressure to break complicated operations into separate functions that work in isolation. Functions that are easier to test are usually easier to understand and recombine in new ways.

Call to action

When we start to fix a bug, we first like to convert it into a (failing) test. This is wonderfully effective at making your goal very concrete: make this test pass. This is basically a special case of a general methodology known as test driven development.

Robust code

If you know that all the major functionality of your package is well covered by the tests, you can confidently make big changes without worrying about accidentally breaking something. This provides a great reality check when you think you've discovered some brilliant new way to simplify your package. Sometimes such "simplifications" fail to account for some important use case and your tests will save you from yourself.

Introducing testthat

This chapter describes how to test your R package using the testthat package (*https://testthat.r-lib.org*).

If you're familiar with frameworks for unit testing in other languages, you should note that there are some fundamental differences with testthat. This is because R is, at heart, more a functional programming language than an object-oriented programming language. For instance, because R's main object-oriented systems (S3 and S4) are based on generic functions (i.e., methods belong to functions not classes), testing approaches built around objects and methods don't make much sense.

testthat 3.0.0 (released 2020-10-31) introduced the idea of an *edition* of testthat, specifically the third edition of testthat, which we refer to as testthat 3e. An edition is a bundle of behaviors that you have to explicitly choose to use, allowing us to make otherwise backward incompatible changes. This is particularly important for testthat since it has a very large number of packages that use it (almost 5,000 at last count). To use testthat 3e, you must have a version of testthat >= 3.0.0 *and* explicitly opt-in to the third edition behaviors. This allows testthat to continue to evolve and improve without breaking historical packages that are in a rather passive maintenance phase. You can learn more in the testthat 3e article (*https://oreil.ly/NUmqk*) and the blog post "Upgrading to testthat edition 3" (*https://oreil.ly/1RPSO*).

We recommend testthat 3e for all new packages, and we recommend updating existing, actively maintained packages to use testthat 3e. Unless we say otherwise, this chapter describes testthat 3e.

Test Mechanics and Workflow

To use testthat, there is some one-time setup, followed by the ongoing creation of new tests. To get the promised benefits of formal testing, it is important that you run these tests regularly.

Initial Setup

To set up your package to use testthat, run:

```
usethis::use_testthat(3)
```

This will:

1. Create a *tests/testthat/* directory.

2. Add testthat to the `Suggests` field in the *DESCRIPTION* and specify testthat 3e in the `Config/testthat/edition` field. The affected `DESCRIPTION` fields might look like:

   ```
   Suggests: testthat (>= 3.0.0)
   Config/testthat/edition: 3
   ```

3. Create a file *tests/testthat.R* that runs all your tests when R CMD check runs (see "check() and R CMD check" on page 60). For a package named "pkg" the contents of this file will be something like:

   ```
   library(testthat)
   library(pkg)

   test_check("pkg")
   ```

This initial setup is usually something you do once per package. However, even in a package that already uses testthat, it is safe to run `use_testthat(3)` when you're ready to opt-in to testthat 3e.

Do not edit *tests/testthat.R*! It is run during R CMD `check` (and, therefore, dev tools::`check()`), but is not used in most other test-running scenarios (such as dev tools::`test()` or devtools::`test_active_file()`). If you want to do something that affects all of your tests, there is almost always a better way than modifying the boilerplate *tests/testthat.R* script. This chapter details many different ways to make objects and logic available during testing.

Create a Test

As you define functions in your package, in the files below *R/*, you add the corresponding tests to *.R* files in *tests/testthat/*. We strongly recommend that the organization of test files matches the organization of *R/* files, discussed in "Organize Functions Into Files" on page 83—the `foofy()` function (and its friends and helpers) should be defined in *R/foofy.R* and their tests should live in *tests/testthat/test-foofy.R*.

```
R                                       tests/testthat
└── foofy.R                             └── test-foofy.R
    foofy <- function(...) {...}            test_that("foofy does this", {...})
                                            test_that("foofy does that", {...})
```

Even if you have different conventions for file organization and naming, note that testthat tests *must* live in files below *tests/testthat/* and these filenames *must* begin with *test*. The test filename is displayed in testthat output, which provides helpful context.[1]

usethis offers a helpful pair of functions for creating or toggling between files:

- `usethis::use_r()`
- `usethis::use_test()`

Either one can be called with a file (base) name, to create a file *de novo* and open it for editing:

```
use_r("foofy")    # creates and opens R/foofy.R
use_test("blarg") # creates and opens tests/testthat/test-blarg.R
```

The `use_r()` / `use_test()` duo has some convenience features that make them "just work" in many common situations:

- When determining the target file, they can deal with the presence or absence of the *.R* extension and the *test-* prefix.
  - Equivalent: `use_r("foofy.R")`, `use_r("foofy")`
  - Equivalent: `use_test("test-blarg.R")`, `use_test("blarg.R")`, `use_test("blarg")`
- If the target file already exists, it is opened for editing. Otherwise, the target is created and then opened for editing.

---

1 The legacy function `testthat::context()` is superseded now and its use in new or actively maintained code is discouraged. In testthat 3e, `context()` is formally deprecated; you should just remove it. Once you adopt an intentional, synchronized approach to the organization of files below *R/* and *tests/testthat/*, the necessary contextual information is right there in the filename, rendering the legacy `context()` superfluous.

RStudio

If *R/foofy.R* is the active file in your source editor, you can even call `use_test()` with no arguments! The target test file can be inferred: if you're editing *R/foofy.R*, you probably want to work on the companion test file, *tests/testthat/test-foofy.R*. If it doesn't exist yet, it is created and, either way, the test file is opened for editing. This all works the other way around also. If you're editing *tests/testthat/test-foofy.R*, a call to `use_r()` (optionally, creates and) opens *R/foofy.R*.

Bottom line: `use_r()`/`use_test()` are handy for initially creating these file pairs and, later, for shifting your attention from one to the other.

When `use_test()` creates a new test file, it inserts an example test:

```
test_that("multiplication works", {
  expect_equal(2 * 2, 4)
})
```

You will replace this with your own description and logic, but it's a nice reminder of the basic form:

- A test file holds one or more `test_that()` tests.
- Each test describes what it's testing, e.g., "multiplication works."
- Each test has one or more expectations, e.g., `expect_equal(2 * 2, 4)`.

In the following sections and chapters, we go into much more detail about how to test your own functions.

Run Tests

Depending on where you are in the development cycle, you'll run your tests at various scales. When you are rapidly iterating on a function, you might work at the level of individual tests. As the code settles down, you'll run entire test files and eventually the entire test suite.

Micro-Iteration

This is the interactive phase where you initiate and refine a function and its tests in tandem. Here you will run `devtools::load_all()` often, and then execute individual expectations or whole tests interactively in the console. Note that `load_all()` attaches testthat, so it puts you in the perfect position to test drive your functions and to execute individual tests and expectations:

```
# tweak the foofy() function and re-load it
devtools::load_all()

# interactively explore and refine expectations and tests
expect_equal(foofy(...), EXPECTED_FOOFY_OUTPUT)

testthat("foofy does good things", {...})
```

Mezzo-Iteration

As one file's-worth of functions and their associated tests start to shape up,
you will want to execute the entire file of associated tests, perhaps with `test
that::test_file()`:

```
testthat::test_file("tests/testthat/test-foofy.R")
```

RStudio

In RStudio, you have a couple shortcuts for running a single test
file.

If the target test file is the active file, you can use the "Run Tests"
button in the upper-right corner of the source editor.

There is also a useful function, `devtools::test_active_file()`. It
infers the target test file from the active file and, similar to how
`use_r()` and `use_test()` work, it works regardless of whether the
active file is a test file or a companion *R/*.R* file. You can invoke
this via "Run a test file" in the Addins menu. However, for heavy
users (like us!), we recommend binding this to a keyboard shortcut
(*https://oreil.ly/26n4o*); we use Ctrl/Cmd-T.

Macro-Iteration

As you near the completion of a new feature or bug fix, you will want to run the
entire test suite.

Most frequently, you'll do this with `devtools::test()`:

```
devtools::test()
```

Then eventually, as part of R CMD check with `devtools::check()`:

```
devtools::check()
```

RStudio

devtools::test() is mapped to Ctrl/Cmd-Shift-T.

devtools::check() is mapped to Ctrl/Cmd-Shift-E.

The output of devtools::test() looks like this:

```
devtools::test()
i Loading usethis
i Testing usethis
✓ | F W S  OK | Context
✓ |          1 | addin [0.1s]
✓ |          6 | badge [0.5s]
    ...
✓ |         27 | github-actions [4.9s]
    ...
✓ |         44 | write [0.6s]

══ Results ═══════════════════════════════════════
Duration: 31.3 s

── Skipped tests ─────────────────────────────────
• Not on GitHub Actions, Travis, or Appveyor (3)

[ FAIL 1 | WARN 0 | SKIP 3 | PASS 728 ]
```

Test failure is reported like this:

```
Failure (test-release.R:108:3): get_release_data() works if no file found
res$Version (`actual`) not equal to "0.0.0.9000" (`expected`).

`actual`:   "0.0.0.1234"
`expected`: "0.0.0.9000"
```

Each failure gives a description of the test (e.g., "get_release_data() works if no file found"), its location (e.g., test-release.R:108:3), and the reason for the failure (e.g., "res$Version (actual) not equal to "0.0.0.9000" (expected)").

The idea is that you'll modify your code (either the functions defined below *R/* or the tests in *tests/testthat/*) until all tests are passing.

Test Organization

A test file lives in *tests/testthat/*. Its name must start with *test*. We will inspect and execute a test file from the stringr package.

But first, for the purposes of rendering this book, we must attach stringr and testthat. Note that in real-life test-running situations, this is taken care of by your package development tooling:

- During interactive development, `devtools::load_all()` makes testthat and the package-under-development available (both its exported and unexported functions).

- During arms' length test execution, this is taken care of by `devtools::test_active_file()`, `devtools::test()`, and *tests/testthat.R*.

 Your test files should not include these `library()` calls. We also explicitly request testthat edition 3, but in a real package this will be declared in *DESCRIPTION*:

```
library(testthat)
library(stringr)
local_edition(3)
```

Here are the contents of *tests/testthat/test-dup.r* from stringr:

```
test_that("basic duplication works", {
  expect_equal(str_dup("a", 3), "aaa")
  expect_equal(str_dup("abc", 2), "abcabc")
expect_equal(str_dup(c("a", "b"), 2), c("aa", "bb"))
expect_equal(str_dup(c("a", "b"), c(2, 3)), c("aa", "bbb"))
})
#> Test passed 😸
test_that("0 duplicates equals empty string", {
expect_equal(str_dup("a", 0), "")
expect_equal(str_dup(c("a", "b"), 0), rep("", 2))
})
#> Test passed 🎉
test_that("uses tidyverse recycling rules", {
expect_error(str_dup(1:2, 1:3), class = "vctrs_error_incompatible_size")
})
#> Test passed 🎉
```

This file shows a typical mix of tests:

- "basic duplication works" tests typical usage of `str_dup()`.
- "0 duplicates equals empty string" probes a specific edge case.
- "uses tidyverse recycling rules" checks that malformed input results in a specific kind of error.

Tests are organized hierarchically: *expectations* are grouped into *tests*, which are organized in *files*:

- A *file* holds multiple related tests. In this example, the file *tests/testthat/test-dup.r* has all of the tests for the code in *R/dup.r*.

- A *test* groups together multiple expectations to test the output from a simple function, a range of possibilities for a single parameter from a more complicated function, or tightly related functionality from across multiple functions. This is why they are sometimes called *unit* tests. Each test should cover a single unit of functionality. A test is created with `test_that(desc, code)`.

  It's common to write the description (`desc`) to create something that reads naturally, e.g., `test_that("basic duplication works", { ... })`. A test failure report includes this description, which is why you want a concise statement of the test's purpose, e.g., a specific behavior.

- An *expectation* is the atom of testing. It describes the expected result of a computation: Does it have the right value and right class? Does it produce an error when it should? An expectation automates visual checking of results in the console. Expectations are functions that start with `expect_`.

You want to arrange things such that, when a test fails, you'll know what's wrong and where in your code to look for the problem. This motivates all our recommendations regarding file organization, file naming, and the test description. Finally, try to avoid putting too many expectations in one test—it's better to have more smaller tests than fewer larger tests.

Expectations

An expectation is the finest level of testing. It makes a binary assertion about whether or not an object has the properties you expect. This object is usually the return value from a function in your package.

All expectations have a similar structure:

- They start with `expect_`.
- They have two main arguments: the first is the actual result, the second is what you expect.
- If the actual and expected results don't agree, testthat throws an error.
- Some expectations have additional arguments that control the finer points of comparing an actual and expected result.

While you'll normally put expectations inside tests inside files, you can also run them directly. This makes it easy to explore expectations interactively. There are more than 40 expectations in the testthat package, which can be explored in testthat's reference index (*https://testthat.r-lib.org/reference/index.html*). We're going to cover only the most important expectations here.

Testing for Equality

expect_equal() checks for equality, with some reasonable amount of numeric tolerance:

```
expect_equal(10, 10)
expect_equal(10, 10L)
expect_equal(10, 10 + 1e-7)
expect_equal(10, 11)
#> Error: 10 (`actual`) not equal to 11 (`expected`).
#>
#>    `actual`: 10
#> `expected`: 11
```

If you want to test for exact equivalence, use expect_identical():

```
expect_equal(10, 10 + 1e-7)
expect_identical(10, 10 + 1e-7)
#> Error: 10 (`actual`) not identical to 10 + 1e-07 (`expected`).
#>
#>    `actual`: 10.0000000
#> `expected`: 10.0000001

expect_equal(2, 2L)
expect_identical(2, 2L)
#> Error: 2 (`actual`) not identical to 2L (`expected`).
#>
#> `actual` is a double vector (2)
#> `expected` is an integer vector (2)
```

Testing Errors

Use expect_error() to check whether an expression throws an error. It's the most important expectation in a trio that also includes expect_warning() and expect_message(). We're going to emphasize errors here, but most of this also applies to warnings and messages.

Usually you care about two things when testing an error:

- Does the code fail? Specifically, does it fail for the right reason?
- Does the accompanying message make sense to the human who needs to deal with the error?

The entry-level solution is to expect a specific type of condition:

```
1 / "a"
#> Error in 1/"a": non-numeric argument to binary operator
expect_error(1 / "a")

log(-1)
#> Warning in log(-1): NaNs produced
```

```
#> [1] NaN
expect_warning(log(-1))
```

This is a bit dangerous, though, especially when testing an error. There are lots of ways for code to fail! Consider the following test:

```
expect_error(str_duq(1:2, 1:3))
```

This expectation is intended to test the recycling behavior of str_dup(). But, due to a typo, it tests behavior of a nonexistent function, str_duq(). The code throws an error and, therefore, the test passes, but for the *wrong reason*. Due to the typo, the actual error thrown is about not being able to find the str_duq() function:

```
str_duq(1:2, 1:3)
#> Error in str_duq(1:2, 1:3): could not find function "str_duq"
```

Historically, the best defense against this was to assert that the condition message matches a certain regular expression, via the second argument, regexp:

```
expect_error(1 / "a", "non-numeric argument")
expect_warning(log(-1), "NaNs produced")
```

This does, in fact, force our typo problem to the surface:

```
expect_error(str_duq(1:2, 1:3), "recycle")
#> Error in str_duq(1:2, 1:3): could not find function "str_duq"
```

Recent developments in both base R and rlang make it increasingly likely that conditions are signaled with a *class*, which provides a better basis for creating precise expectations. That is exactly what you've already seen in this stringr example. This is what the class argument is for:

```
# fails, error has wrong class
expect_error(str_duq(1:2, 1:3), class = "vctrs_error_incompatible_size")
#> Error in str_duq(1:2, 1:3): could not find function "str_duq"

# passes, error has expected class
expect_error(str_dup(1:2, 1:3), class = "vctrs_error_incompatible_size")
```

If you have the choice, express your expectation in terms of the condition's class, instead of its message. Often this is under your control, i.e., if your package signals the condition. If the condition originates from base R or another package, proceed with caution. This is often a good reminder to reconsider the wisdom of testing a condition that is not fully under your control in the first place.

To check for the *absence* of an error, warning, or message, use expect_no_error():

```
expect_no_error(1 / 2)
```

Of course, this is functionally equivalent to simply executing 1 / 2 inside a test, but some developers find the explicit expectation expressive.

If you genuinely care about the condition's message, testthat 3e's snapshot tests are the best approach, which we describe next.

Snapshot Tests

Sometimes it's difficult or awkward to describe an expected result with code. Snapshot tests are a great solution to this problem, and this is one of the main innovations in testthat 3e. The basic idea is that you record the expected result in a separate, human-readable file. Going forward, testthat alerts you when a newly computed result differs from the previously recorded snapshot. Snapshot tests are particularly suited to monitoring your package's user interface, such as its informational messages and errors. Other use cases include testing images or other complicated objects.

We'll illustrate snapshot tests using the waldo package. Under the hood, testthat 3e uses waldo to do the heavy lifting of "actual versus expected" comparisons, so it's good for you to know a bit about waldo anyway. One of waldo's main design goals is to present differences in a clear and actionable manner, as opposed to a frustrating declaration that "this differs from that and I know exactly how, but I won't tell you." Therefore, the formatting of output from waldo::compare() is very intentional and is well-suited to a snapshot test. The binary outcome of TRUE (actual == expected) versus FALSE (actual != expected) is fairly easy to check and could get its own test. Here we're concerned with writing a test to ensure that differences are reported to the user in the intended way.

waldo uses a few different layouts for showing diffs, depending on various conditions. Here we deliberately constrain the width, in order to trigger a side-by-side layout.[2] (We talk more about the withr package elsewhere in the book, including later in this chapter.)

```
withr::with_options(
  list(width = 20),
  waldo::compare(c("X", letters), c(letters, "X"))
)
#>      old | new
#> [1] "X" -
#> [2] "a" | "a" [1]
#> [3] "b" | "b" [2]
#> [4] "c" | "c" [3]
#>
#>      old | new
```

---

2 The actual waldo test that inspires this example targets an unexported helper function that produces the desired layout. But this example uses an exported waldo function for simplicity.

```
#> [25] "x" | "x" [24]
#> [26] "y" | "y" [25]
#> [27] "z" | "z" [26]
#>          - "X" [27]
```

The two primary inputs differ at two locations: once at the start and once at the end. This layout presents both of these, with some surrounding context, which helps the reader orient themselves.

Here's how this would look as a snapshot test:

```
test_that("side-by-side diffs work", {
  withr::local_options(width = 20)
  expect_snapshot(
    waldo::compare(c("X", letters), c(letters, "X"))
  )
})
```

If you execute expect_snapshot() or a test containing expect_snapshot() interactively, you'll see this:

```
Can't compare snapshot to reference when testing interactively
i Run `devtools::test()` or `testthat::test_file()` to see changes
```

followed by a preview of the snapshot output.

This reminds you that snapshot tests function only when executed noninteractively, i.e., while running an entire test file or the entire test suite. This applies both to recording snapshots and to checking them.

The first time this test is executed via devtools::test() or similar, you'll see something like this (assume the test is in *tests/testthat/test-diff.R*):

```
— Warning (test-diff.R:63:3): side-by-side diffs work ————————————
Adding new snapshot:
Code
  waldo::compare(c(
    "X", letters), c(
    letters, "X"))
Output
      old | new
  [1] "X" -
  [2] "a" | "a" [1]
  [3] "b" | "b" [2]
  [4] "c" | "c" [3]

       old | new
  [25] "x" | "x" [24]
  [26] "y" | "y" [25]
  [27] "z" | "z" [26]
           - "X" [27]
```

There is always a warning upon initial snapshot creation. The snapshot is added to *tests/testthat/_snaps/diff.md*, under the heading "side-by-side diffs work," which comes from the test's description. The snapshot looks exactly like what a user sees interactively in the console, which is the experience we want to check for. The snapshot file is *also* very readable, which is pleasant for the package developer. This readability extends to snapshot changes, i.e., when examining Git diffs and reviewing pull requests on GitHub, which helps you keep tabs on your user interface. Going forward, as long as your package continues to recapitulate the expected snapshot, this test will pass.

If you've written a lot of conventional unit tests, you can appreciate how well-suited snapshot tests are for this use case. If we were forced to inline the expected output in the test file, there would be a great deal of quoting, escaping, and newline management. Ironically, with conventional expectations, the output you expect your user to see tends to get obscured by a heavy layer of syntactical noise.

What about when a snapshot test fails? Let's imagine a hypothetical internal change where the default labels switch from "old" and "new" to "OLD" and "NEW." Here's how this snapshot test would react:

```
— Failure (test-diff.R:63:3): side-by-side diffs work————————————————
Snapshot of code has changed:
old[3:15] vs new[3:15]
    "     \"X\", letters), c("
    "     letters, \"X\"))"
    "Output"
-   "        old | new      "
+   "        OLD | NEW      "
    " [1] \"X\" -            "
    " [2] \"a\" | \"a\" [1]"
    " [3] \"b\" | \"b\" [2]"
    " [4] \"c\" | \"c\" [3]"
    "  "
-   "        old | new      "
+   "        OLD | NEW      "
and 3 more ...

* Run `snapshot_accept('diff')` to accept the change
* Run `snapshot_review('diff')` to interactively review the change
```

This diff is presented more effectively in most real-world usage, e.g., in the console, by a Git client, or via a Shiny app (see the next paragraph). But even this plain-text version highlights the changes quite clearly. Each of the two loci of change is indicated with a pair of lines marked with - and +, showing how the snapshot has changed.

You can call `testthat::snapshot_review('diff')` to review changes locally in a Shiny app, which lets you skip or accept individual snapshots. Or, if all changes are intentional and expected, you can go straight to `testthat::snap`

---

shot_accept('diff'). Once you've resynchronized your actual output and the snapshots on file, your tests will pass once again. In real life, snapshot tests are a great way to stay informed about changes to your package's user interface, due to your own internal changes or due to changes in your dependencies or even R itself.

expect_snapshot() has a few arguments worth knowing about:

cran = FALSE

By default, snapshot tests are skipped if it looks like the tests are running on CRAN's servers. This reflects the typical intent of snapshot tests, which is to proactively monitor user interface but not to check for correctness, which presumably is the job of other unit tests that are not skipped. In typical usage, a snapshot change is something the developer will want to know about, but it does not signal an actual defect.

error = FALSE

By default, snapshot code is *not* allowed to throw an error. See expect_error(), described previously, for one approach to testing errors. But sometimes you want to assess "Does this error message make sense to a human?" and having it laid out in context in a snapshot is a great way to see it with fresh eyes. Specify error = TRUE in this case:

```
expect_snapshot(error = TRUE,
  str_dup(1:2, 1:3)
)
```

transform

Sometimes a snapshot contains volatile, insignificant elements, such as a temporary filepath or a timestamp. The transform argument accepts a function, presumably written by you, to remove or replace such changeable text. Another use of transform is to scrub sensitive information from the snapshot.

variant

Sometimes snapshots reflect the ambient conditions, such as the operating system or the version of R or one of your dependencies, and you need a different snapshot for each variant. This is an experimental and somewhat advanced feature, so if you can arrange things to use a single snapshot, you probably should.

In typical usage, testthat will take care of managing the snapshot files below *tests/ testthat/_snaps/*. This happens in the normal course of you running your tests and, perhaps, calling testthat::snapshot_accept().

Shortcuts for Other Common Patterns

We conclude this section with a few more expectations that come up frequently. But remember that testthat has many more prebuilt expectations (*https://testthat.r-lib.org/reference/index.html*) than we can demonstrate here.

Several expectations can be described as "shortcuts," i.e., they streamline a pattern that comes up often enough to deserve its own wrapper:

- `expect_match(object, regexp, ...)` is a shortcut that wraps `grepl(pattern = regexp, x = object, ...)`. It matches a character vector input against a regular expression `regexp`. The optional `all` argument controls whether all elements or just one element needs to match. Read the `expect_match()` documentation to see how additional arguments, like `ignore.case = FALSE` or `fixed = TRUE`, can be passed down to `grepl()`:

  ```
  string <- "Testing is fun!"

  expect_match(string, "Testing")

  # Fails, match is case-sensitive
  expect_match(string, "testing")
  #> Error: `string` does not match "testing".
  #> Actual value: "Testing is fun!"

  # Passes because additional arguments are passed to grepl():
  expect_match(string, "testing", ignore.case = TRUE)
  ```

- `expect_length(object, n)` is a shortcut for `expect_equal(length(object), n)`.

- `expect_setequal(x, y)` tests that every element of x occurs in y, and that every element of y occurs in x. But it won't fail if x and y happen to have their elements in a different order.

- `expect_s3_class()` and `expect_s4_class()` check that an object inherit() from a specified class. `expect_type()` checks the typeof() an object:

  ```
  model <- lm(mpg ~ wt, data = mtcars)
  expect_s3_class(model, "lm")
  expect_s3_class(model, "glm")
  #> Error: `model` inherits from 'lm' not 'glm'.
  ```

`expect_true()` and `expect_false()` are useful catchalls if none of the other expectations does what you need.

## CHAPTER 14

Designing Your Test Suite

This chapter runs a few small bits of testthat code, so we must do some setup that is not necessary in organic testthat usage.

 Your test files should not include these library() calls. We also explicitly request testthat edition 3, but in a real package this will be declared in *DESCRIPTION*:

```
library(testthat)
local_edition(3)
```

What to Test

> Whenever you are tempted to type something into a print statement or a debugger expression, write it as a test instead. — Martin Fowler

There is a fine balance to writing tests. Each test that you write makes your code less likely to change inadvertently; but it also can make it harder to change your code on purpose. It's hard to give good general advice about writing tests, but you might find these points helpful:

- Focus on testing the external interface to your functions—if you test the internal interface, then it's harder to change the implementation in the future because as well as modifying the code, you'll also need to update all the tests.

- Strive to test each behavior in one and only one test. Then if that behavior later changes you need to update only a single test.

- Avoid testing simple code that you're confident will work. Instead focus your time on code that you're not sure about, is fragile, or has complicated interdependencies. That said, we often find we make the most mistakes when we falsely assume that the problem is simple and doesn't need any tests.

- Always write a test when you discover a bug. You may find it helpful to adopt the test-first philosophy. There you always start by writing the tests, and then write the code that makes them pass. This reflects an important problem-solving strategy: start by establishing your success criteria, or how you know if you've solved the problem.

Test Coverage

Another concrete way to direct your test writing efforts is to examine your test coverage. The covr package (*https://covr.r-lib.org*) can be used to determine which lines of your package's source code are (or are not!) executed when the test suite is run. This is most often presented as a percentage. Generally speaking, the higher the better.

In some technical sense, 100% test coverage is the goal; however, this is rarely achieved in practice and that's often OK. Going from 90% or 99% coverage to 100% is not always the best use of your development time and energy. In many cases, that last 10% or 1% often requires some awkward gymnastics to cover. Sometimes this forces you to introduce mocking or some other new complexity. Don't sacrifice the maintainability of your test suite in the name of covering some weird edge case that hasn't yet proven to be a problem. Also remember that not every line of code or every function is equally likely to harbor bugs. Focus your testing energy on code that is tricky, based on your expert opinion and any empirical evidence you've accumulated about bug hot spots.

We use covr regularly, in two different ways:

Local, interactive use

We mostly use `devtools::test_coverage_active_file()` and `devtools::test_coverage()` for exploring the coverage of an individual file or the whole package, respectively.

Automatic, remote use via GitHub Actions (GHA)

We cover continuous integration and GHA more thoroughly in Chapter 20, but we will at least mention here that `usethis::use_github_action("test-coverage")` configures a GHA workflow that constantly monitors your test coverage. Test coverage can be an especially helpful metric when evaluating a pull request (either your own or from an external contributor). A proposed change that is well-covered by tests is less risky to merge.

High-Level Principles for Testing

In later sections, we offer concrete strategies for how to handle common testing dilemmas in R. Here we lay out the high-level principles that underpin these recommendations:

- A test should ideally be self-sufficient and self-contained.
- The interactive workflow is important, because you will mostly interact with your tests when they are failing.
- It's more important that test code be obvious, e.g., as DRY ("don't repeat yourself") as possible.
- However, the interactive workflow shouldn't "leak" into and undermine the test suite.

Writing good tests for a code base often feels more challenging than writing the code in the first place. This can come as a bit of a shock when you're new to package development and you might be concerned that you're doing it wrong. Don't worry, you're not! Testing presents many unique challenges and maneuvers, which tend to get much less air time in programming communities than strategies for writing the "main code," i.e., the stuff below *R/*. As a result, it requires more deliberate effort to develop your skills and taste around testing.

Many of the packages maintained by our team violate some of the advice you'll find here. There are (at least) two reasons for that:

- testthat has been evolving for more than twelve years, and this chapter reflects the cumulative lessons learned from that experience. The tests in many packages have been in place for a long time and reflect typical practices from different eras and different maintainers.
- These aren't hard-and-fast rules but are, rather, guidelines. There will always be specific situations where it makes sense to bend the rule.

This chapter can't address all possible testing situations, but hopefully these guidelines will help your future decision-making.

Self-Sufficient Tests

> All tests should strive to be hermetic: a test should contain all of the information necessary to set up, execute, and tear down its environment. Tests should assume as little as possible about the outside environment ....
>
> —From the book *Software Engineering at Google*, Chapter 11 (https://oreil.ly/9U0Nc)

Recall this advice found in "Respect the R Landscape" on page 91, which covers your package's "main code," i.e., everything below *R/*:

> The *.R* files below *R/* should consist almost entirely of function definitions. Any other top-level code is suspicious and should be carefully reviewed for possible conversion into a function.

We have analogous advice for your test files:

> The *test-*.R* files below *tests/testthat/* should consist almost entirely of calls to test_that(). Any other top-level code is suspicious and should be carefully considered for relocation into calls to test_that() or to other files that get special treatment inside an R package or from testthat.

Eliminating (or at least minimizing) top-level code outside of test_that() will have the beneficial effect of making your tests more hermetic. This is basically the testing analogue of the general programming advice that it's wise to avoid unstructured sharing of state.

Logic at the top-level of a test file has an awkward scope: objects or functions defined here have what you might call "test file scope," if the definitions appear before the first call to test_that(). If top-level code is interleaved between test_that() calls, you can even create "partial test file scope."

While writing tests, it can feel convenient to rely on these file-scoped objects, especially early in the life of a test suite, e.g., when each test file fits on one screen. But we find that implicitly relying on objects in a test's parent environment tends to make a test suite harder to understand and maintain over time.

Consider a test file with top-level code sprinkled around it, outside of test_that():

```
dat <- data.frame(x = c("a", "b", "c"), y = c(1, 2, 3))

skip_if(today_is_a_monday())

test_that("foofy() does this", {
  expect_equal(foofy(dat), ...)
})

dat2 <- data.frame(x = c("x", "y", "z"), y = c(4, 5, 6))

skip_on_os("windows")

test_that("foofy2() does that", {
  expect_snapshot(foofy2(dat, dat2))
})
```

We recommend relocating file-scoped logic to either a narrower scope or to a broader scope. Here's what it would look like to use a narrow scope, i.e., to inline everything inside test_that() calls:

```
test_that("foofy() does this", {
  skip_if(today_is_a_monday())

  dat <- data.frame(x = c("a", "b", "c"), y = c(1, 2, 3))

  expect_equal(foofy(dat), ...)
})
test_that("foofy() does that", {
  skip_if(today_is_a_monday())
  skip_on_os("windows")

  dat <- data.frame(x = c("a", "b", "c"), y = c(1, 2, 3))
  dat2 <- data.frame(x = c("x", "y", "z"), y = c(4, 5, 6))

  expect_snapshot(foofy(dat, dat2))
})
```

Later, we will discuss techniques for moving file-scoped logic to a broader scope.

Self-Contained Tests

Each test_that() test has its own execution environment, which makes it somewhat self-contained. For example, an R object you create inside a test does not exist after the test exits:

```
exists("thingy")
#> [1] FALSE

test_that("thingy exists", {
  thingy <- "thingy"
  expect_true(exists(thingy))
})
#> Test passed 

exists("thingy")
#> [1] FALSE
```

The thingy object lives and dies entirely within the confines of test_that(). However, testthat doesn't know how to clean up after actions that affect other aspects of the R landscape:

The filesystem
    Creating and deleting files, changing the working directory, etc.

The search path
    Changing the search path with library() or attach().

Global options, graphical parameters, and environment variables
    modifying options, parameters, and environment variables with options(), par(), and Sys.setenv().

Watch how calls like library(), options(), and Sys.setenv() have a persistent effect *after* a test, even when they are executed inside test_that():

```
grep("jsonlite", search(), value = TRUE)
#> character(0)
getOption("opt_whatever")
#> NULL
Sys.getenv("envvar_whatever")
#> [1] ""

test_that("landscape changes leak outside the test", {
  library(jsonlite)
  options(opt_whatever = "whatever")
  Sys.setenv(envvar_whatever = "whatever")

  expect_match(search(), "jsonlite", all = FALSE)
  expect_equal(getOption("opt_whatever"), "whatever")
  expect_equal(Sys.getenv("envvar_whatever"), "whatever")
})
#> Test passed 

grep("jsonlite", search(), value = TRUE)
#> [1] "package:jsonlite"
getOption("opt_whatever")
#> [1] "whatever"
Sys.getenv("envvar_whatever")
#> [1] "whatever"
```

These changes to the landscape persist even beyond the current test file, i.e., they carry over into all subsequent test files.

If it's easy to avoid making such changes in your test code, that is the best strategy! But if it's unavoidable, then you have to make sure that you clean up after yourself. This mindset is very similar to one we advocated in "Respect the R Landscape" on page 91, when discussing how to design well-mannered functions.

We like to use the withr package (*https://withr.r-lib.org*) to make temporary changes in global state, because it automatically captures the initial state and arranges the eventual restoration. You've already seen an example of its usage, when we explored snapshot tests:

```
test_that("side-by-side diffs work", {
  withr::local_options(width = 20) # <-- (°_°) look here!
  expect_snapshot(
    waldo::compare(c("X", letters), c(letters, "X"))
  )
})
```

This test requires the display width to be set at 20 columns, which is considerably less than the default width. `withr::local_options(width = 20)` sets the `width` option to 20 and, at the end of the test, restores the option to its original value. withr is also pleasant to use during interactive development: deferred actions are still captured on the global environment and can be executed explicitly via `withr::deferred_run()` or implicitly by restarting R.

We recommend including withr in `Suggests`, if you're only going to use it in your tests, or in `Imports`, if you also use it below *R/*. Call withr functions as we do in the previous snippet, e.g., like `withr::local_whatever()`, in either case. See "Whether to Import or Depend" on page 152 and "In Test Code" on page 162 for more.

 The easiest way to add a package to *DESCRIPTION* is with, e.g., `usethis::use_package("withr", type = "Suggests")`. For tidyverse packages, withr is considered a "free dependency," i.e., the tidyverse uses withr so extensively that we don't hesitate to use it whenever it would be useful.

withr has a large set of pre-implemented `local_*()` / `with_*()` functions that should handle most of your testing needs, so check there before you write your own. If nothing exists that meets your need, `withr::defer()` is the general way to schedule some action at the end of a test.[1]

Here's how we would fix the problems in the previous example using withr. *Behind the scenes, we reversed the landscape changes, so we can try this again:*

```
grep("jsonlite", search(), value = TRUE)
#> character(0)
getOption("opt_whatever")
#> NULL
Sys.getenv("envvar_whatever")
#> [1] ""

test_that("withr makes landscape changes local to a test", {
  withr::local_package("jsonlite")
  withr::local_options(opt_whatever = "whatever")
  withr::local_envvar(envvar_whatever = "whatever")

  expect_match(search(), "jsonlite", all = FALSE)
  expect_equal(getOption("opt_whatever"), "whatever")
  expect_equal(Sys.getenv("envvar_whatever"), "whatever")
```

---

1 Base R's `on.exit()` is another alternative, but it requires more from you. You need to capture the original state and write the restoration code yourself. Also remember to do `on.exit(..., add = TRUE)` if there's *any* chance a second `on.exit()` call could be added in the test. You probably also want to default to `after = FALSE`.

```
})
#> Test passed 🐢

grep("jsonlite", search(), value = TRUE)
#> character(0)
getOption("opt_whatever")
#> NULL
Sys.getenv("envvar_whatever")
#> [1] ""
```

testthat leans heavily on withr to make test execution environments as reproducible and self-contained as possible. In testthat 3e, `testthat::local_reproducible_out put()` is implicitly part of each `test_that()` test:

```
test_that("something specific happens", {
  local_reproducible_output() # <-- this happens implicitly

  # your test code, which might be sensitive to ambient conditions, such as
  # display width or the number of supported colors
})
```

`local_reproducible_output()` temporarily sets various options and environment variables to values favorable for testing, e.g., it suppresses colored output, turns off fancy quotes, sets the console width, and sets `LC_COLLATE = "C"`. Usually, you can just passively enjoy the benefits of `local_reproducible_output()`. But you may want to call it explicitly when replicating test results interactively or if you want to override the default settings in a specific test.

Plan for Test Failure

We regret to inform you that most of the quality time you spend with your tests will be when they are inexplicably failing:

> In its purest form, automating testing consists of three activities: writing tests, running tests, and *reacting to test failures*....

> Remember that tests are often revisited only when something breaks. When you are called to fix a broken test that you have never seen before, you will be thankful someone took the time to make it easy to understand. Code is read far more than it is written, so make sure you write the test you'd like to read!

> —From the book *Software Engineering at Google*, Chapter 11 (https://oreil.ly/9U0Nc)

Most of us don't work on a code base the size of Google. But even in a team of one, tests that you wrote six months ago might as well have been written by someone else. Especially when they are failing.

When we do reverse dependency checks, often involving hundreds or thousands of CRAN packages, we have to inspect test failures to determine if changes in our packages are to blame. As a result, we regularly engage with failing tests in other people's packages, which leaves us with lots of opinions about practices that create unnecessary testing pain.

Test troubleshooting nirvana looks like this: in a fresh R session, you can do devtools::load_all() and immediately run an individual test or walk through it line-by-line. There is no need to hunt around for setup code that has to be run manually first, that is found elsewhere in the test file or perhaps in a different file altogether. Test-related code that lives in an unconventional location causes extra self-inflicted pain when you least need it.

Consider this extreme and abstract example of a test that is difficult to troubleshoot due to implicit dependencies on free-range code:

```
# dozens or hundreds of lines of top-level code, interspersed with other tests,
# which you must read and selectively execute

test_that("f() works", {
  x <- function_from_some_dependency(object_with_unknown_origin)
  expect_equal(f(x), 2.5)
})
```

This test is much easier to drop in on if dependencies are invoked in the normal way, i.e., via ::, and test objects are created inline:

```
# dozens or hundreds of lines of self-sufficient, self-contained tests,
# all of which you can safely ignore!

test_that("f() works", {
  useful_thing <- ...
  x <- somePkg::someFunction(useful_thing)
  expect_equal(f(x), 2.5)
})
```

This test is self-sufficient. The code inside { ... } explicitly creates any necessary objects or conditions and makes explicit calls to any helper functions. This test doesn't rely on objects or dependencies that happen to be ambiently available.

Self-sufficient, self-contained tests are a win-win: it is literally safer to design tests this way, and it also makes tests much easier for humans to troubleshoot later.

Repetition Is OK

One obvious consequence of our suggestion to minimize code with "file scope" is that your tests will probably have some repetition. And that's OK! We're going to make the controversial recommendation that you tolerate a fair amount of duplication in test code, i.e., you can relax some of your DRY ("don't repeat yourself") tendencies.

> Keep the reader in your test function. Good production code is well-factored; good test code is obvious. ... think about what will make the problem obvious when a test fails.
>
> —From the blog post "Why Good Developers Write Bad Unit Tests" (https://oreil.ly/9kDBl)

Here's a toy example to make things concrete:

```
test_that("multiplication works", {
  useful_thing <- 3
  expect_equal(2 * useful_thing, 6)
})
#> Test passed
```

```
test_that("subtraction works", {
  useful_thing <- 3
  expect_equal(5 - useful_thing, 2)
})
#> Test passed
```

In real life, `useful_thing` is usually a more complicated object that somehow feels burdensome to instantiate. Notice how `useful_thing <- 3` appears in more than one place. Conventional wisdom says we should DRY this code out. It's tempting to just move `useful_thing`'s definition outside of the tests:

```
useful_thing <- 3

test_that("multiplication works", {
  expect_equal(2 * useful_thing, 6)
})
#> Test passed
```

```
test_that("subtraction works", {
  expect_equal(5 - useful_thing, 2)
})
#> Test passed
```

But we really do think the first form, with the repetition, is often the better choice.

At this point, many readers might be thinking "but the code I might have to repeat is much longer than 1 line!" Later we describe the use of test fixtures. This can often reduce complicated situations back to something that resembles this simple example.

Remove Tension Between Interactive and Automated Testing

Your test code will be executed in two different settings:

- Interactive test development and maintenance, which includes tasks like:
  — Initial test creation
  — Modifying tests to adapt to change
  — Debugging test failure
- Automated test runs, which are accomplished with functions such as:
  — Single file: `devtools::test_active_file()`, `testthat::test_file()`
  — Whole package: `devtools::test()`, `devtools::check()`

Automated testing of your whole package is what takes priority. This is ultimately the whole point of your tests. However, the interactive experience is clearly important for the humans doing this work. Therefore it's important to find a pleasant workflow but also to ensure that you don't rig anything for interactive convenience that actually compromises the health of the test suite.

These two modes of test-running should not be in conflict with each other. If you perceive tension between these two modes, this can indicate you're not taking full advantage of some of testthat's features and the way it's designed to work with `devtools::load_all()`.

When working on your tests, use `load_all()`, just like you do when working below *R/*. By default, `load_all()` does all of these things:

- Simulates rebuilding, reinstalling, and reloading your package.
- Makes everything in your package's namespace available, including unexported functions and objects and anything you've imported from another package.
- Attaches testthat, i.e., does `library(testthat)`.
- Runs test helper files, i.e., executes *test/testthat/helper.R* (more on that soon).

This eliminates the need for any `library()` calls below *tests/testthat/*, for the vast majority of R packages. Any instance of `library(testthat)` is clearly no longer necessary. Likewise, any instance of attaching one of your dependencies via `library(somePkg)` is unnecessary. In your tests, if you need to call functions from somePkg, do it just as you do below *R/*. If you have imported the function into your namespace, use `fun()`. If you have not, use `somePkg::fun()`. It's fair to say that `library(somePkg)` in the tests should be about as rare as taking a dependency via `Depends`, i.e., there is almost always a better alternative.

Unnecessary calls to library(somePkg) in test files have a real downside, because they actually change the R landscape. library() alters the search path. This means the circumstances under which you are testing may not necessarily reflect the circumstances under which your package will be used. This makes it easier to create subtle test bugs, which you will have to unravel in the future.

One other function that should almost never appear below *tests/testhat/* is source(). There are several special files with an official role in testthat workflows (see the next section), not to mention the entire R package machinery, that provide better ways to make functions, objects, and other logic available in your tests.

Files Relevant to Testing

Here we review which package files are especially relevant to testing and, more generally, best practices for interacting with the filesystem from your tests.

Hiding in Plain Sight: Files Below R/

The most important functions you'll need to access from your tests are clearly those in your package! Here we're talking about everything that's defined below *R/*. The functions and other objects defined by your package are always available when testing, regardless of whether or not they are exported. For interactive work, devtools::load_all() takes care of this. During automated testing, this is taken care of internally by testthat.

This implies that test helpers can absolutely be defined below *R/* and used freely in your tests. It might make sense to gather such helpers in a clearly marked file, such as one of these:

```
.
├── ...
└── R
    ├── ...
    ├── test-helpers.R
    ├── test-utils.R
    ├── testthat.R
    ├── utils-testing.R
    └── ...
```

For example, the dbplyr package uses *R/testthat.R* (*https://oreil.ly/dN0iQ*) to define a couple of helpers to facilitate comparisons and expectations involving tbl objects, which is used to represent database tables:

```
compare_tbl <- function(x, y, label = NULL, expected.label = NULL) {
  testthat::expect_equal(
    arrange(collect(x), dplyr::across(everything())),
    arrange(collect(y), dplyr::across(everything())),
    label = label,
```

```
    expected.label = expected.label
  )
}

expect_equal_tbls <- function(results, ref = NULL, ...) {
  # code that gets things ready ...

  for (i in seq_along(rest)) {
    compare_tbl(
      rest[[i]], ref,
      label = names(rest)[[i]],
      expected.label = ref_name
    )
  }

  invisible(TRUE)
}
```

tests/testthat.R

Recall the initial testthat setup described in "Test Mechanics and Workflow" on page 186. The standard *tests/testthat.R* file looks like this:

```
library(testthat)
library(pkg)

test_check("pkg")
```

We repeat the advice to not edit *tests/testthat.R*. It is run during R CMD check (and, therefore, devtools::check()) but is not used in most other test-running scenarios (such as devtools::test() or devtools::test_active_file() or during interactive development). Do not attach your dependencies here with library(). Call them in your tests in the same manner as you do below *R/* ("In Test Code" on page 161, and "In Test Code" on page 162).

testthat Helper Files

Another type of file that is always executed by load_all() and at the beginning of automated testing is a helper file, defined as any file below *tests/testthat/* that begins with *helper*. Helper files are a mighty weapon in the battle to eliminate code floating around at the top-level of test files. Helper files are a prime example of what we mean when we recommend moving such code into a broader scope. Objects or functions defined in a helper file are available to all of your tests.

If you have just one such file, you should probably name it *helper.R*. If you organize your helpers into multiple files, you could include a suffix with additional info. Here are examples of how such files might look:

```
├── ...
└── tests
    ├── testthat
    │   ├── helper.R
    │   ├── helper-blah.R
    │   ├── helper-foo.R
    │   ├── test-foofy.R
    │   └── (more test files)
    └── testthat.R
```

Many developers use helper files to define custom test helper functions, which we describe in detail in Chapter 15. Compared to defining helpers below *R/*, some people find that *tests/testthat/helper.R* makes it more clear that these utilities are specifically for testing the package. This location also feels more natural if your helpers rely on testthat functions. For example, usethis (*https://oreil.ly/582ha*) and vroom (*https://oreil.ly/z3vzz*) both have fairly extensive *tests/testthat/helper.R* files that define many custom test helpers. Here are two very simple usethis helpers that check that the currently active project (usually an ephemeral test project) has a specific file or folder:

```
expect_proj_file <- function(...) expect_true(file_exists(proj_path(...)))
expect_proj_dir <- function(...) expect_true(dir_exists(proj_path(...)))
```

A helper file is also a good location for setup code that is needed for its side effects. This is a case where *tests/testthat/helper.R* is clearly more appropriate than a file below *R/*. For example, in an API-wrapping package, *helper.R* is a good place to (attempt to) authenticate with the testing credentials.[2]

testthat Setup Files

testthat has one more special file type: setup files, defined as any file below *test/testthat/* that begins with *setup*. Here's an example of how that might look:

```
├── ...
└── tests
    ├── testthat
    │   ├── helper.R
    │   ├── setup.R
    │   ├── test-foofy.R
    │   └── (more test files)
    └── testthat.R
```

---

2 googledrive does this in *https://oreil.ly/fWgjm*.

A setup file is handled almost exactly like a helper file, but with two big differences:

- Setup files are not executed by `devtools::load_all()`.
- Setup files often contain the corresponding teardown code.

Setup files are good for global test setup that is tailored for test execution in non-interactive or remote environments. For example, you might turn off behavior that's aimed at an interactive user, such as messaging or writing to the clipboard.

If any of your setup should be reversed after test execution, you should also include the necessary teardown code in *setup.R*.[3] We recommend maintaining teardown code alongside the setup code, in *setup.R*, because this makes it easier to ensure they stay in sync. The artificial environment `teardown_env()` exists as a magical handle to use in `withr::defer()` and `withr::local_*()` / `withr::with_*()`.

Here's a *setup.R* example from the reprex package, where we turn off clipboard and HTML preview functionality during testing:

```
op <- options(reprex.clipboard = FALSE, reprex.html_preview = FALSE)

withr::defer(options(op), teardown_env())
```

Since we are just modifying options here, we can be even more concise and use the prebuilt function `withr::local_options()` and pass `teardown_env()` as the `.local_envir`:

```
withr::local_options(
  list(reprex.clipboard = FALSE, reprex.html_preview = FALSE),
  .local_envir = teardown_env()
)
```

Files Ignored by testthat

testthat automatically executes files only where these are both true:

- File is a direct child of *tests/testthat/*
- Filename starts with one of the specific strings:
  - *helper*
  - *setup*
  - *test*

---

3 A legacy approach (which still works but is no longer recommended) is to put teardown code in *tests/testthat/teardown.R*.

It is fine to have other files or directories in *tests/testthat/*, but testthat won't automatically do anything with them (other than the *_snaps* directory, which holds snapshots).

Storing Test Data

Many packages contain files that hold test data. Where should these be stored? The best location is somewhere below *tests/testthat/*, often in a subdirectory, to keep things neat. Following is an example, where *useful_thing1.rds* and *useful_thing2.rds* hold objects used in the test files:

```
.
├── ...
└── tests
    ├── testthat
    │   ├── fixtures
    │   │   ├── make-useful-things.R
    │   │   ├── useful_thing1.rds
    │   │   └── useful_thing2.rds
    │   ├── helper.R
    │   ├── setup.R
    │   └── (all the test files)
    └── testthat.R
```

Then, in your tests, use testthat::test_path() to build a robust filepath to such files:

```
test_that("foofy() does this", {
  useful_thing <- readRDS(test_path("fixtures", "useful_thing1.rds"))
  # ...
})
```

testthat::test_path() is extremely handy, because it produces the correct path in the two important modes of test execution:

- Interactive test development and maintenance, where the working directory is presumably set to the top-level of the package.

- Automated testing, where the working directory is usually set to something below *tests/*.

Where to Write Files During Testing

If it's easy to avoid writing files from your tests, that is definitely the best plan. But many times you really must write files.

You should write files only inside the session temp directory. Do not write into your package's *tests/* directory. Do not write into the current working directory. Do not write into the user's home directory. Even though you are writing into the session temp directory, you should still clean up after yourself, i.e., delete any files you've written.

Most package developers don't want to hear this, because it sounds like a hassle. But it's not that burdensome once you get familiar with a few techniques and build some new habits. A high level of filesystem discipline also eliminates various testing bugs and will absolutely make your CRAN life run more smoothly.

This test is from roxygen2 and demonstrates everything we recommend:

```
test_that("can read from file name with utf-8 path", {
  path <- withr::local_tempfile(
    pattern = "Universit\u00e0-",
    lines = c("#' @include foo.R", NULL)
  )
  expect_equal(find_includes(path), "foo.R")
})
```

`withr::local_tempfile()` creates a file within the session temp directory whose lifetime is tied to the "local" environment—in this case, the execution environment of an individual test. It is a wrapper around `base::tempfile()` and passes, e.g., the `pattern` argument through, so you have some control over the filename. You can optionally provide `lines` to populate the file at creation time or you can write to the file in all the usual ways in subsequent steps. Finally, with no special effort on your part, the temporary file will automatically be deleted at the end of the test.

Sometimes you need even more control over the filename. In that case, you can use `withr::local_tempdir()` to create a self-deleting temporary directory and write intentionally named files inside this directory.

## CHAPTER 15

Advanced Testing Techniques

This chapter runs a few small bits of testthat code, so we must do some setup that is not necessary in organic testthat usage.

 Your test files should not include these library() calls. We also explicitly request testthat edition 3, but in a real package this will be declared in *DESCRIPTION*:

```
library(testthat)
local_edition(3)
```

Test Fixtures

When it's not practical to make your test entirely self-sufficient, prefer making the necessary object, logic, or conditions available in a structured, explicit way. There's a preexisting term for this in software engineering: a *test fixture*.

> A test fixture is something used to consistently test some item, device, or piece of software. — Wikipedia

The main idea is that we need to make it as easy and obvious as possible to arrange the world into a state that is conducive to testing. We describe several specific solutions to this problem:

- Put repeated code in a constructor-type helper function. Memoise it, if construction is demonstrably slow.

- If the repeated code has side effects, write a custom local_*() function to do what's needed and clean up afterwards.

- If the preceding approaches are too slow or awkward and the thing you need is fairly stable, save it as a static file and load it.

Create useful_things with a Helper Function

Is it fiddly to create a `useful_thing`? Does it take several lines of code but not much time or memory? In that case, write a helper function to create a `useful_thing` on-demand:

```
new_useful_thing <- function() {
  # your fiddly code to create a useful_thing goes here
}
```

and call that helper in the affected tests:

```
test_that("foofy() does this", {
  useful_thing1 <- new_useful_thing()
  expect_equal(foofy(useful_thing1, x = "this"), EXPECTED_FOOFY_OUTPUT)
})

test_that("foofy() does that", {
  useful_thing2 <- new_useful_thing()
  expect_equal(foofy(useful_thing2, x = "that"), EXPECTED_FOOFY_OUTPUT)
})
```

Where should the `new_useful_thing()` helper be defined? This comes back to what we outlined in "Files Relevant to Testing" on page 212. Test helpers can be defined below *R/*, just like any other internal utility in your package. Another popular location is in a test helper file, e.g., *tests/testthat/helper.R*. A key feature of both options is that the helpers are made available to you during interactive maintenance via `devtools::load_all()`.

If it's fiddly *and* costly to create a `useful_thing`, your helper function could even use memoisation to avoid unnecessary recomputation. Once you have a helper like `new_useful_thing()`, you often discover that it has uses beyond testing, e.g., behind-the-scenes in a vignette. Sometimes you even realize you should just define it below *R/* and export and document it, so you can use it freely in documentation and tests.

Create (and Destroy) a Local useful_thing

So far, our example of a `useful_thing` was a regular R object, which is cleaned up automatically at the end of each test. What if the creation of a `useful_thing` has a side effect on the local filesystem, on a remote resource, R session options, environment variables, or the like? Then your helper function should create a `useful_thing` *and clean up afterwards*. Instead of a simple `new_useful_thing()` constructor, you'll write a customized function in the style of withr's `local_*()` functions:

```
local_useful_thing <- function(..., env = parent.frame()) {
  # your fiddly code to create a useful_thing goes here
  withr::defer(
    # your fiddly code to clean up after a useful_thing goes here
    envir = env
  )
}
```

Use it in your tests like this:

```
test_that("foofy() does this", {
  useful_thing1 <- local_useful_thing()
  expect_equal(foofy(useful_thing1, x = "this"), EXPECTED_FOOFY_OUTPUT)
})

test_that("foofy() does that", {
  useful_thing2 <- local_useful_thing()
  expect_equal(foofy(useful_thing2, x = "that"), EXPECTED_FOOFY_OUTPUT)
})
```

Where should the `local_useful_thing()` helper be defined? All the advice given previously for `new_useful_thing()` applies: define it below *R/* or in a test helper file.

To learn more about writing custom helpers like `local_useful_thing()`, see the testthat vignette on test fixtures (*https://oreil.ly/2nJm-*).

Store a Concrete useful_thing Persistently

If a `useful_thing` is costly to create, in terms of time or memory, maybe you don't actually need to re-create it for each test run. You could make the `useful_thing` once, store it as a static test fixture, and load it in the tests that need it. Here's a sketch of how this could look:

```
test_that("foofy() does this", {
  useful_thing1 <- readRDS(test_path("fixtures", "useful_thing1.rds"))
  expect_equal(foofy(useful_thing1, x = "this"), EXPECTED_FOOFY_OUTPUT)
})

test_that("foofy() does that", {
  useful_thing2 <- readRDS(test_path("fixtures", "useful_thing2.rds"))
  expect_equal(foofy(useful_thing2, x = "that"), EXPECTED_FOOFY_OUTPUT)
})
```

Now we can revisit a file listing from earlier, which addressed exactly this scenario:

```
.
├── ...
└── tests
    ├── testthat
    │   ├── fixtures
    │   │   ├── make-useful-things.R
    │   │   ├── useful_thing1.rds
```

```
|   |        └── useful_thing2.rds
|   ├── helper.R
|   ├── setup.R
|   └── (all the test files)
└── testthat.R
```

This shows static test files stored in *tests/testthat/fixtures/*, but also notice the companion R script, *make-useful-things.R*. From data analysis, we all know there is no such thing as a script that is run only once. Refinement and iteration are inevitable. This also holds true for test objects like *useful_thing1.rds*. We highly recommend saving the R code used to create your test objects, so that they can be re-created as needed.

Building Your Own Testing Tools

Let's return to the topic of duplication in your test code. We've encouraged you to have a higher tolerance for repetition in test code, in the name of making your tests obvious. But there's still a limit to how much repetition to tolerate. We've covered techniques such as loading static objects with `test_path()`, writing a constructor like `new_useful_thing()`, or implementing a test fixture like `local_useful_thing()`. There are even more types of test helpers that can be useful in certain situations.

Helper Defined Inside a Test

Consider this test for the `str_trunc()` function in stringr:

```
# from stringr (hypothetically)
test_that("truncations work for all sides", {
  expect_equal(
    str_trunc("This string is moderately long", width = 20, side = "right"),
    "This string is mo..."
  )
  expect_equal(
    str_trunc("This string is moderately long", width = 20, side = "left"),
    "...s moderately long"
  )
  expect_equal(
    str_trunc("This string is moderately long", width = 20, side = "center"),
    "This stri...ely long"
  )
})
```

There's a lot of repetition here, which increases the chance of copy/paste errors and generally makes your eyes glaze over. Sometimes it's nice to create a hyper-local helper, *inside the test*. Here's how the test actually looks in stringr:

```
# from stringr (actually)
test_that("truncations work for all sides", {
```

```
    trunc <- function(direction) str_trunc(
      "This string is moderately long",
      direction,
      width = 20
    )

    expect_equal(trunc("right"),    "This string is mo...")
    expect_equal(trunc("left"),     "...s moderately long")
    expect_equal(trunc("center"),   "This stri...ely long")
})
```

A hyper-local helper like `trunc()` is particularly useful when it allows you to fit all the important business for each expectation on one line. Then your expectations can be read almost like a table of actual versus expected, for a set of related use cases. In the preceding example, it's very easy to watch the result change as we truncate the input from the right, left, and in the center.

Note that this technique should be used in extreme moderation. A helper like `trunc()` is yet another place where you can introduce a bug, so it's best to keep such helpers extremely short and simple.

Custom Expectations

If a more complicated helper feels necessary, it's a good time to reflect on why that is. If it's fussy to get into position to *test* a function, that could be a sign that it's also fussy to *use* that function. Do you need to refactor it? If the function seems sound, then you probably need to use a more formal helper, defined outside of any individual test, as described earlier.

One specific type of helper you might want to create is a custom expectation. Here are two very simple ones from usethis:

```
expect_usethis_error <- function(...) {
  expect_error(..., class = "usethis_error")
}

expect_proj_file <- function(...) {
  expect_true(file_exists(proj_path(...)))
}
```

`expect_usethis_error()` checks that an error has the `"usethis_error"` class. `expect_proj_file()` is a simple wrapper around `file_exists()` that searches for the file in the current project. These are very simple functions, but the sheer amount of repetition and the expressiveness of their names makes them feel justified.

It is somewhat involved to make a proper custom expectation, i.e., one that behaves like the expectations built into testthat. We refer you to the "Custom expectations" vignette (*https://oreil.ly/G1vfN*) if you wish to learn more about that.

Finally, it can be handy to know that testthat makes specific information available when it's running:

- The environment variable TESTTHAT is set to "true". testthat::is_testing() is a shortcut:

  ```
  is_testing <- function() {
    Sys.getenv("TESTTHAT")
  }
  ```

- The package-under-test is available as the environment variable TESTTHAT_PKG and testthat::testing_package() is a shortcut:

  ```
  testing_package <- function() {
    Sys.getenv("TESTTHAT_PKG")
  }
  ```

In some situations, you may want to exploit this information without taking a runtime dependency on testthat. In that case, just inline the source of these functions directly into your package.

When Testing Gets Hard

Despite all the techniques we've covered so far, there remain situations where it still feels very difficult to write tests. In this section, we review more ways to deal with challenging situations:

- Skipping a test in certain situations
- Mocking an external service
- Dealing with secrets

Skipping a Test

Sometimes it's impossible to perform a test—you may not have an internet connection or you may not have access to the necessary credentials. Unfortunately, another likely reason follows from this simple rule: the more platforms you use to test your code, the more likely it is that you won't be able to run all of your tests, all of the time. In short, there are times when, instead of getting a failure, you just want to skip a test.

testthat::skip()

Here we use testthat::skip() to write a hypothetical custom skipper, skip_if_no_api():

```
skip_if_no_api() <- function() {
  if (api_unavailable()) {
```

```
      skip("API not available")
    }
  }

  test_that("foo api returns bar when given baz", {
    skip_if_no_api()
    ...
  })
```

`skip_if_no_api()` is yet another example of a test helper, and the advice already given about where to define it applies here too.

`skip()`s and the associated reasons are reported inline as tests are executed and are also indicated clearly in the summary:

```
devtools::test()
#> i Loading abcde
#> i Testing abcde
#> ✓ | F W S  OK | Context
#> ✓ |         2 | blarg
#> ✓ |      1  2 | foofy
#> ─────────────────────────────────────────────────────────
#> Skip (test-foofy.R:6:3): foo api returns bar when given baz
#> Reason: API not available
#> ─────────────────────────────────────────────────────────
#> ✓ |         0 | yo
#> ══ Results ═══════════════════════════════════════════════
#>> ── Skipped tests ───────────────────────────────────────
#> • API not available (1)
#>
#> [ FAIL 0 | WARN 0 | SKIP 1 | PASS 4 ]
#>
#> 
```

Something like `skip_if_no_api()` is likely to appear many times in your test suite. This is another occasion where it is tempting to DRY things out, by hoisting the `skip()` to the top-level of the file. However, we still lean toward calling `skip_if_no_api()` in each test where it's needed:

```
# we prefer this:
test_that("foo api returns bar when given baz", {
  skip_if_no_api()
  ...
})

test_that("foo api returns an errors when given qux", {
  skip_if_no_api()
  ...
})

# over this:
skip_if_no_api()
```

```
test_that("foo api returns bar when given baz", {...})

test_that("foo api returns an errors when given qux", {...})
```

Within the realm of top-level code in test files, having a skip() at the very beginning of a test file is one of the more benign situations. But once a test file does not fit entirely on your screen, it creates an implicit yet easy-to-miss connection between the skip() and individual tests.

Built-In skip() functions

Similar to testthat's built-in expectations, a family of skip() functions anticipates some common situations. These functions often relieve you of the need to write a custom skipper. Here are some examples of the most useful skip() functions:

```
test_that("foo api returns bar when given baz", {
  skip_if(api_unavailable(), "API not available")
  ...
})
test_that("foo api returns bar when given baz", {
  skip_if_not(api_available(), "API not available")
  ...
})

skip_if_not_installed("sp")
skip_if_not_installed("stringi", "1.2.2")

skip_if_offline()
skip_on_cran()
skip_on_os("windows")
```

Dangers of skipping

One challenge with skips is that they are currently completely invisible in CI—if you automatically skip too many tests, it's easy to fool yourself that all your tests are passing when in fact they're just being skipped! In an ideal world, your CI/CD would make it easy to see how many tests are being skipped and how that changes over time.

It's a good practice to regularly dig into the R CMD check results, especially on CI, and make sure the skips are as you expect. But this tends to be something you have to learn through experience.

Mocking

In the practice known as mocking, we replace something that's complicated or unreliable or out of our control with something simpler, that's fully within our control. Usually we are mocking an external service, such as a REST API, or a function that reports something about session state, such as whether the session is interactive.

The classic application of mocking is in the context of a package that wraps an external API. In order to test your functions, technically you need to make a live call to that API to get a response, which you then process. But what if that API requires authentication or what if it's somewhat flaky and has occasional downtime? It can be more productive to just *pretend* to call the API but, instead, to test the code under your control by processing a prerecorded response from the actual API.

Our main advice about mocking is to avoid it if you can. This is not an indictment of mocking, just a realistic assessment that mocking introduces new complexity that is not always justified by the payoffs.

Since most R packages do not need full-fledged mocking, we do not cover it here. Instead we'll point you to the packages that represent the state-of-the-art for mocking in R today:

- mockery (*https://github.com/r-lib/mockery*)
- mockr (*https://krlmlr.github.io/mockr*)
- httptest (*https://enpiar.com/r/httptest*)
- httptest2 (*https://enpiar.com/httptest2*)
- webfakes (*https://webfakes.r-lib.org*)

Note also that, at the time of writing, it seems likely that the testthat package will reintroduce some mocking capabilities (after getting out of the mocking business once already). Version v3.1.7 has two new experimental functions: `testthat::with_mocked_bindings()` and `testthat::local_mocked_bindings()`.

Secrets

Another common challenge for packages that wrap an external service is the need to manage credentials. Specifically, it is likely that you will need to provide a set of test credentials to fully test your package.

Our main advice here is to design your package so that large parts of it can be tested without live, authenticated access to the external service.

Of course, you still want to be able to test your package against the actual service that it wraps, in environments that support secure environment variables. Since this is also a very specialized topic, we won't go into more detail here. Instead we refer you to the Wrapping APIs vignette (*https://oreil.ly/NYmvQ*) in the httr2 package, which offers substantial support for secret management.

Special Considerations for CRAN Packages

CRAN runs R CMD check on all contributed packages, both upon submission and on a regular basis after acceptance. This check includes, but is not limited to, your testthat tests. We discuss the general challenge of preparing your package to face all of CRAN's check "flavors" in "CRAN Check Flavors and Related Services" on page 329. Here we focus on CRAN-specific considerations for your test suite.

When a package runs afoul of the CRAN Repository Policy (*https://oreil.ly/NbCYF*), the test suite is very often the culprit (although not always). If your package is destined for CRAN, this should influence how you write your tests and how (or whether) they will be run on CRAN.

Skip a Test

If a specific test simply isn't appropriate to be run by CRAN, include skip_on_cran() at the very start:

```
test_that("some long-running thing works", {
  skip_on_cran()
  # test code that can potentially take "a while" to run
})
```

Under the hood, skip_on_cran() consults the NOT_CRAN environment variable. Such a test will run when NOT_CRAN has been explicitly defined as "true". This variable is set by devtools and testthat, allowing those tests to run in environments where you expect success (and where you can tolerate and troubleshoot occasional failure).

In particular, the GitHub Actions workflows that we recommend in "GitHub Actions" on page 298 *will* run tests with NOT_CRAN = "true". For certain types of functionality, there is no practical way to test it on CRAN and your own checks, on GitHub Actions or an equivalent continuous integration service, are your best method of quality assurance.

In rare cases it makes sense to maintain tests outside of your package altogether. The tidymodels team uses this strategy for integration-type tests of their whole ecosystem that would be impossible to host inside an individual CRAN package.

Speed

Your tests need to run relatively quickly—ideally, less than a minute, in total. Use skip_on_cran() in a test that is unavoidably long-running.

Reproducibility

Be careful about testing things that are likely to be variable on CRAN machines. It's risky to test how long something takes (because CRAN machines are often heavily loaded) or to test parallel code (because CRAN runs multiple package tests in parallel, multiple cores will not always be available). Numerical precision can also vary across platforms, so use `expect_equal()` unless you have a specific reason for using `expect_identical()`.

Flaky Tests

Due to the scale at which CRAN checks packages, there is basically no latitude for a test that's "just flaky," i.e., sometimes fails for incidental reasons. CRAN does not process your package's test results the way you do, where you can inspect each failure and exercise some human judgment about how concerning it is.

It's probably a good idea to eliminate flaky tests, just for your own sake! But if you have valuable, well-written tests that are prone to occasional nuisance failure, definitely put `skip_on_cran()` at the start.

The classic example is any test that accesses a website or web API. Given that any web resource in the world will experience occasional downtime, it's best to not let such tests run on CRAN. The CRAN Repository Policy says:

> Packages which use Internet resources should fail gracefully with an informative message if the resource is not available or has changed (and not give a check warning nor error).

Often making such a failure "graceful" would run counter to the behavior you actually want in practice, i.e., you would want your user to get an error if their request fails. This is why it is usually more practical to test such functionality elsewhere.

Recall that snapshot tests (see Chapter 13), by default, are also skipped on CRAN. You typically use such tests to monitor, e.g., how various informational messages look. Slight changes in message formatting are something you want to be alerted to, but they do not indicate a major defect in your package. This is the motivation for the default `skip_on_cran()` behavior of snapshot tests.

Finally, flaky tests cause problems for the maintainers of your dependencies. When the packages you depend on are updated, CRAN runs `R CMD check` on all reverse dependencies, including your package. If your package has flaky tests, your package can be the reason another package does not clear CRAN's incoming checks and can delay its release.

Process and Filesystem Hygiene

In "Where to Write Files During Testing" on page 216, we urged you to write only into the session temp directory and to clean up after yourself. This practice makes your test suite much more maintainable and predictable. For packages that are (or aspire to be) on CRAN, this is absolutely required per the CRAN repository policy:

> Packages should not write in the user's home filespace (including clipboards), nor anywhere else on the file system apart from the R session's temporary directory (or during installation in the location pointed to by TMPDIR: and such usage should be cleaned up).... Limited exceptions may be allowed in interactive sessions if the package obtains confirmation from the user.

Similarly, you should make an effort to be hygienic with respect to any processes you launch:

> Packages should not start external software (such as PDF viewers or browsers) during examples or tests unless that specific instance of the software is explicitly closed afterwards.

Accessing the clipboard is the perfect storm that potentially runs afoul of both of these guidelines, as the clipboard is considered part of the user's home filespace and, on Linux, can launch an external process (e.g., xsel or xclip). Therefore it is best to turn off any clipboard functionality in your tests (and to ensure that, during authentic usage, your user is clearly opting-in to that).

## PART V

Documentation

## CHAPTER 16

Function Documentation

In this chapter, you'll learn about function documentation, which users access with ?somefunction or help("somefunction"). Base R provides a standard way of documenting a package where each function is documented in a *topic*, an *.Rd* file ("R documentation") in the *man/* directory. *.Rd* files use a custom syntax, loosely based on LaTeX, and can be rendered to HTML, plain text, or PDF, as needed, for viewing in different contexts.

In the devtools ecosystem, we don't edit *.Rd* files directly. Instead, we include specially formatted "roxygen comments" above the source code for each function.[1] Then we use the roxygen2 package (*https://oreil.ly/ygJdd*) to generate the *.Rd* files from these special comments.[2] There are a few advantages to using roxygen2:

- Code and documentation are co-located. When you modify your code, it's easy to remember to also update your documentation.

- You can use markdown, rather than having to learn a one-off markup language that only applies to *.Rd* files. In addition to formatting, the automatic hyperlinking functionality makes it much, much easier to create richly linked documentation.

- There's a lot of *.Rd* boilerplate that's automated away.

- roxygen2 provides a number of tools for sharing content across documentation topics and even between topics and vignettes.

---

1 The name "roxygen" is a nod to the Doxygen documentation generator, which inspired the development of an R package named roxygen. Then that original concept was rebooted as roxygen2, similar to ggplot2.

2 The *NAMESPACE* file is also generated from these roxygen comments. Or, rather, it *can* be and that is the preferred devtools workflow (see "NAMESPACE Workflow" on page 156).

In this chapter we'll focus on documenting functions, but the same ideas apply to documenting datasets (see "Documenting Datasets" on page 104), classes and generics, and packages. You can learn more about those important topics in `vignette("rd-other", package = "roxygen2")`.

roxygen2 Basics

To get started, we'll work through the basic roxygen2 workflow and discuss the overall structure of roxygen2 comments, which are organized into blocks and tags. We also highlight the biggest wins of using markdown with roxygen2.

The Documentation Workflow

Unlike with testthat, there's no obvious opening move to declare that you're going to use roxygen2 for documentation. That's because the use of roxygen2 is purely a matter of your development workflow. It has no effect on, e.g., how a package gets checked or built. We think the roxygen approach is the best way to generate your *.Rd* files, but officially R only cares about the files themselves, not how they came to be.

Your documentation workflow truly begins when you start to add roxygen comments above your functions. Roxygen comment lines always start with `#'` , the usual `#` for a comment, followed immediately by a single quote `'`:

```
#' Add together two numbers
#'
#' @param x A number.
#' @param y A number.
#' @returns A numeric vector.
#' @examples
#' add(1, 1)
#' add(10, 1)
add <- function(x, y) {
  x + y
}
```

RStudio

Usually you write your function first, then its documentation. Once the function definition exists, put your cursor somewhere in it and do Code > Insert Roxygen Skeleton to get a great head start on the roxygen comment.

Once you have at least one roxygen comment, run `devtools::document()` to generate (or update) your package's *.Rd* files.[3] Under the hood, this ultimately calls `roxygen2::roxygenise()`. The preceding roxygen block generates a *man/add.Rd* file that looks like this:

```
% Generated by roxygen2: do not edit by hand
% Please edit documentation in R/add.R
\name{add}
\alias{add}
\title{Add together two numbers}
\usage{
add(x, y)
}
\arguments{
\item{x}{A number.}

\item{y}{A number.}
}
\value{
A numeric vector.
}
\description{
Add together two numbers
}
\examples{
add(1, 1)
add(10, 1)
}
```

RStudio

You can also run `devtools::document()` with the keyboard shortcut Ctrl/Cmd-Shift-D or via the Build menu or pane.

If you've used LaTeX before, this should look vaguely familiar since the *.Rd* format is loosely based on LaTeX. If you are interested in the *.Rd* format, you can read more in *Writing R Extensions* (*https://oreil.ly/gIwJP*). But generally you'll never need to look at *.Rd* files, except to commit them to your package's Git repository.

---

3 Running `devtools::document()` also affects another field in *DESCRIPTION*, which looks like this: Roxygen Note: 7.2.1. This records which version of roxygen2 was last used in a package, which makes it easier for devtools (and its underlying packages) to make an intelligent guess about when to re-`document()` a package and when to leave well enough alone. In a collaborative setting, this also reduces nuisance changes to the *.Rd* files, by making the relevant roxygen2 version highly visible.

How does this *.Rd* file correspond to the documentation you see in R? When you run ?add, help("add"), or example("add"), R looks for an *.Rd* file containing \alias{add}. It then parses the file, converts it into HTML, and displays it. Figure 16-1 shows how this help topic would look in RStudio.

---

add {rvest}   R Documentation

Add together two numbers

Description

Add together two numbers

Usage

`add(x, y)`

Arguments

x A number
y A number

Value

The sum of x and y

Examples

```
add(1, 1)
add(10, 1)
```

Figure 16-1. Help topic rendered to HTML

R CMD check Warning

You should document all exported functions and datasets. Otherwise, you'll get this warning from R CMD check:

```
W  checking for missing documentation entries (614ms)
   Undocumented code objects:
     'somefunction'
   Undocumented data sets:
     'somedata'
   All user-level objects in a package should have
   documentation entries.
```

Conversely, you probably don't want to document unexported functions. If you want to use roxygen comments for internal documentation, include the @noRd tag to suppress the creation of the *.Rd* file.

This is also a good time to explain something you may have noticed in your *DESCRIPTION* file:

```
Roxygen: list(markdown = TRUE)
```

devtools/usethis includes this by default when initiating a *DESCRIPTION* file and it gives roxygen2 a heads-up that your package uses markdown syntax in its roxygen comments.[4]

The default help-seeking process looks inside *installed* packages, so to see your package's documentation during development, devtools overrides the usual help functions with modified versions that know to consult the current *source* package. To activate these overrides, you'll need to run `devtools::load_all()` at least once. If it feels like your edits to the roxygen comments aren't having an effect, double-check that you have actually regenerated the *.Rd* files with `devtools::document()` and that you've loaded your package. When you call `?function`, you should see "Rendering development documentation …"

To summarize, there are four steps in the basic roxygen2 workflow:

1. Add roxygen2 comments to your *.R* files.
2. Run `devtools::document()` or press Ctrl/Cmd-Shift-D to convert roxygen2 comments to *.Rd* files.
3. Preview documentation with `?function`.
4. Rinse and repeat until the documentation looks the way you want.

roxygen2 Comments, Blocks, and Tags

Now that you understand the basic workflow, we'll go into more detail about the syntax. roxygen2 comments start with `#'` and all the roxygen2 comments preceding a function are collectively called a *block*. Blocks are broken up by *tags*, which look like `@tagName tagValue`, and the content of a tag extends from the end of the tag name to the start of the next tag.[5] A block can contain text before the first tag, which is called the *introduction*. By default, each block generates a single documentation *topic*, i.e., a single *.Rd* file[6] in the *man/* directory.

---

4 This is part of the explanation promised in "Custom Fields" on page 134, where we also clarify that, with our current conventions, this field should really be called `Config/Needs/roxygen`, instead of `Roxygen`. We highly recommend that you use markdown in all new packages and that you migrate older-but-actively-maintained packages to markdown syntax. In this case, you can call `usethis::use_roxygen_md()` to update *DESCRIPTION* and get a reminder about the roxygen2md package, which can help with conversion.

5 Or the end of the block, if it's the last tag.

6 The name of the file is automatically derived from the object you're documenting.

Throughout this chapter we'll show you roxygen2 comments from real tidyverse packages, focusing on stringr (*https://stringr.tidyverse.org*), since the functions there tend to be fairly straightforward, leading to documentation that's understandable with relatively little context. We attach stringr here so that its functions are hyperlinked in the rendered book (more on that in "Key Markdown Features" on page 239):

```
library(stringr)
```

Here's a simple first example—the documentation for str_unique():

```
#' Remove duplicated strings
#'
#' `str_unique()` removes duplicated values, with optional control over
#' how duplication is measured.
#'
#' @param string Input vector. Either a character vector, or something
#'   coercible to one.
#' @param ... Other options used to control matching behavior between duplicate
#'   strings. Passed on to [stringi::stri_opts_collator()].
#' @returns A character vector, usually shorter than `string`.
#' @seealso [unique()], [stringi::stri_unique()] which this function wraps.
#' @examples
#' str_unique(c("a", "b", "c", "b", "a"))
#'
#' # Use ... to pass additional arguments to stri_unique()
#' str_unique(c("motley", "mötley", "pinguino", "pingüino"))
#' str_unique(c("motley", "mötley", "pinguino", "pingüino"), strength = 1)
#' @export
str_unique <- function(string, ...) {
  ...
}
```

Here the introduction includes the title ("Remove duplicated strings") and a basic description of what the function does. The introduction is followed by five tags: two @params, one @returns, one @seealso, one @examples, and one @export.

Note that the block has an intentional line length (typically the same as that used for the surrounding R code) and the second and subsequent lines of the long @param tag are indented, which makes the entire block easier to scan. You can get more roxygen2 style advice in the tidyverse style guide (*https://oreil.ly/Gj0uV*).

RStudio

It can be aggravating to manually manage the line length of roxygen comments, so be sure to try out Code > Reflow Comment (Ctrl/Cmd-Shift-/).

Note also that the order in which tags appear in your roxygen comments (or even in handwritten *.Rd* files) does not dictate the order in rendered documentation. The order of presentation is determined by tooling within base R.

The following sections go into more depth for the most important tags. We start with the introduction, which provides the title, description, and details. Then we cover the inputs (the function arguments), outputs (the return value), and examples. Next we discuss links and cross-references, then finish off with techniques for sharing documentation between topics.

Key Markdown Features

For the most part, general markdown and R Markdown knowledge suffice for taking advantage of markdown in roxygen2. But there are a few pieces of syntax that are so important we want to highlight them here. You'll see these in many of the examples in this chapter:

Backticks for inline code
Use backticks to format a piece of text as code, i.e., in a fixed width font. Example:

```
#' I like `thisfunction()`, because it's great.
```

Square brackets for an auto-linked function
Enclose text like `somefunction()` and `somepackage::somefunction()` in square brackets to get an automatic link to that function's documentation. Be sure to include the trailing parentheses, because it's good style and it causes the function to be formatted as code, i.e., you don't need to add backticks. Example:

```
#' It's obvious that `thisfunction()` is better than
#' [otherpkg::otherfunction()] #' or even our own
#' [olderfunction()].
```

Vignettes
If you refer to a vignette with an inline call to `vignette("some-topic")`, it serves a dual purpose. First, this is literally the R code you would execute to view a vignette locally. But wait, there's more! In many rendered contexts, this automatically becomes a hyperlink to that same vignette in a pkgdown website. Here we use that to link to some very relevant vignettes:[7]

---

7 These calls include an explicit specification of `package = "somepackage"`, since it can't be inferred from context, i.e., the context is a Quarto book, not package documentation.

- vignette("rd-formatting", package = "roxygen2") (*https://oreil.ly/qKlxg*)

- vignette("reuse", package = "roxygen2") (*https://oreil.ly/Y8dvT*)

- vignette("linking", package = "pkgdown") (*https://oreil.ly/OhCQA*)

Lists

Bullet lists break up the dreaded "wall of text" and can make your documentation easier to scan. You can use them in the description of the function or of an argument and also for the return value. It is not necessary to include a blank line before the list, but that is also allowed.

```
#' Best features of `thisfunction()`:
#' * Smells nice
#' * Has good vibes
```

Title, Description, Details

The introduction provides a title, description, and, optionally, details for the function. While it's possible to use explicit tags in the introduction, we usually rely on implicit tags when possible:

- The *title* is taken from the first sentence. It should be written in sentence case, not end in a full stop, and be followed by a blank line. The title is shown in various function indexes (e.g., help(package = "somepackage")) and is what the user will usually see when browsing multiple functions.

- The *description* is taken from the next paragraph. It's shown at the top of documentation and should briefly describe the most important features of the function.

- Additional *details* are anything after the description. Details are optional, but can be any length; they are useful if you want to dig deep into some important aspect of the function. Note that, even though the details come right after the description in the introduction, they appear much later in rendered documentation.

The following sections describe each component in more detail, and then discuss a few useful related tags.

Title

When writing the title, it's useful to think about how it will appear in the reference index. When a user skims the index, how will they know which functions will solve their current problem? This requires thinking about what your functions have in common (which doesn't need to be repeated in every title) and what is unique to that function (which should be highlighted in the title).

When we wrote this chapter, we found the function titles for stringr to be somewhat disappointing. But they provide a useful negative case study:

`str_detect()`
> Detect the presence or absence of a pattern in a string

`str_extract()`
> Extract matching patterns from a string

`str_locate()`
> Locate the position of patterns in a string

`str_match()`
> Extract matched groups from a string

There's a lot of repetition ("pattern," "from a string") and the verb used for the function name is repeated in the title, so if you don't understand the function already, the title seems unlikely to help much. Hopefully we'll have improved those titles by the time you read this!

In contrast, these titles from dplyr are much better:[8]

`mutate()`
> Create, modify, and delete columns

`summarize()`
> Summarize each group down to one row

`filter()`
> Keep rows that match a condition

`select()`
> Keep or drop columns using their names and types

`arrange()`
> Order rows using column values

Here we try to succinctly describe what the function does, making sure to describe whether it affects rows, columns, or groups. We do our best to use synonyms, instead of repeating the function name, to hopefully give folks another chance to understand the intent of the function.

---

8 Like all the examples, these might have changed a bit since we wrote this book, because we're constantly striving to do better. You might compare what's in the book to what we now use, and consider if you think if it's an improvement.

Description

The purpose of the description is to summarize the goal of the function, usually in a single paragraph. This can be challenging for simple functions, because it can feel like you're just repeating the title of the function. Try to find a slightly different wording, if you can. It's OK if this feels a little repetitive; it's often useful for users to see the same thing expressed in two different ways. It's a little extra work, but the extra effort is often worth it. Here's the description for `str_detect()`:

```
#' Detect the presence/absence of a match
#'
#' `str_detect()` returns a logical vector with `TRUE` for each element of
#' `string` that matches `pattern` and `FALSE` otherwise. It's equivalent to
#' `grepl(pattern, string)`.
```

If you want more than one paragraph, you must use an explicit @description tag to prevent the second (and subsequent) paragraphs from being turned into the @details. Here's a two-paragraph @description from `str_view()`:

```
#' View strings and matches
#'
#' @description
#' `str_view()` is used to print the underlying representation of a string and
#' to see how a `pattern` matches.
#'
#' Matches are surrounded by `<>` and unusual whitespace (i.e. all whitespace
#' apart from `" "` and `"\n"`) are surrounded by `{}` and escaped. Where
#' possible, matches and unusual whitespace are coloured blue and `NA`s red.
```

Here's another example from `str_like()`, which has a bullet list in @description:

```
#' Detect a pattern in the same way as `SQL`'s `LIKE` operator
#'
#' @description
#' `str_like()` follows the conventions of the SQL `LIKE` operator:
#'
#' * Must match the entire string.
#' * `&#x5f;` matches a single character (like `.`).
#' * `%` matches any number of characters (like `.*`).
#' * `\%` and `\&#x5f;` match literal `%` and `&#x5f;`.
#' * The match is case insensitive by default.
```

Basically, if you're going to include an empty line in your description, you'll need to use an explicit @description tag.

Finally, it's often particularly hard to write a good description if you've just written the function, because the purpose often seems very obvious. Do your best, and then come back later, when you've forgotten exactly what the function does. Once you've re-derived what the function does, you'll be able to write a better description.

Details

The `@details` are just any additional details or explanation that you think your function needs. Most functions don't need details, but some functions need a lot. If you have a lot of information to convey, it's a good idea to use informative markdown headings to break the details up into manageable sections.[9] Here's an example from dplyr::mutate(). We've elided some of the details to keep this example short, but you should still get a sense of how we used headings to break up the content into skimmable chunks:

```
#' Create, modify, and delete columns
#'
#' `mutate()` creates new columns that are functions of existing variables.
#' It can also modify (if the name is the same as an existing
#' column) and delete columns (by setting their value to `NULL`).
#'
#' @section Useful mutate functions:
#'
#' * [`+`], [`-`], [log()], etc., for their usual mathematical meanings
#'
#' ...
#'
#' @section Grouped tibbles:
#'
#' Because mutating expressions are computed within groups, they may
#' yield different results on grouped tibbles. This will be the case
#' as soon as an aggregating, lagging, or ranking function is
#' involved. Compare this ungrouped mutate:
#'
#' ...
```

This is a good time to remind ourselves that, even though a heading like Useful mutate functions in the previous example comes immediately after the description in the roxygen block, the content appears much later in the rendered documentation. The details (whether they use section headings or not) appear after the function usage, arguments, and return value.

Arguments

For most functions, the bulk of your work will go toward documenting how each argument affects the output of the function. For this purpose, you'll use @param (short for parameter, a synonym of argument) followed by the argument name and a description of its action.

---

9 In older code, you might see the use of @section title: which was used to create sections before roxygen2 had full markdown support. If you've used these in the past, you can now turn them into markdown headings.

The highest priority is to provide a succinct summary of the allowed inputs and what the parameter does. For example, here's how `str_detect()` documents `string`:

```
#' @param string Input vector. Either a character vector, or something
#'   coercible to one.
```

And here are three of the arguments to `str_flatten()`:

```
#' @param collapse String to insert between each piece. Defaults to `""`.
#' @param last Optional string to use in place of the final separator.
#' @param na.rm Remove missing values? If `FALSE` (the default), the result
#'   will be `NA` if any element of `string` is `NA`.
```

Note that `@param collapse` and `@param na.rm` describe their default arguments. This is often a good practice because the function usage (which shows the default values) and the argument description are often quite far apart in the rendered documentation. But there are downsides. The main one is that this duplication means you'll need to make updates in two places if you change the default value; we believe this small amount of extra work is worth it to make the life of the user easier.

If an argument has a fixed set of possible parameters, you should list them. If they're simple, you can just list them in a sentence, like in `str_trim()`:

```
#' @param side Side on which to remove whitespace: `"left"`, `"right"`, or
#'   `"both"` (the default).
```

If they need more explanation, you might use a bulleted list, as in `str_wrap()`:

```
#' @param whitespace_only A boolean.
#'   * `TRUE` (the default): wrapping will only occur at whitespace.
#'   * `FALSE`: can break on any non-word character (e.g. `/`, `-`).
```

The documentation for most arguments will be relatively short, often one or two sentences. But you should take as much space as you need, and you'll see some examples of multiparagraph argument docs shortly.

Multiple Arguments

If the behavior of multiple arguments is tightly coupled, you can document them together by separating the names with commas (with no spaces). For example, x and y are interchangeable in `str_equal()`, so they're documented together:

```
#' @param x,y A pair of character vectors.
```

In `str_sub()`, `start` and `end` define the range of characters to replace. But instead of supplying both, you can use just `start` if you pass in a two-column matrix. So it makes sense to document them together:

```
#' @param start,end A pair of integer vectors defining the range of characters
#'    to extract (inclusive).
#'
#'    Alternatively, instead of a pair of vectors, you can pass a matrix to
#'    `start`. The matrix should have two columns, either labelled `start`
#'    and `end`, or `start` and `length`.
```

In str_wrap(), indent and exdent define the indentation for the first line and all subsequent lines, respectively:

```
#' @param indent,exdent A non-negative integer giving the indent for the
#'    first line (`indent`) and all subsequent lines (`exdent`).
```

Inheriting Arguments

If your package contains many closely related functions, it's common for them to have arguments that share the same name and meaning. It would be both annoying and error prone to copy and paste the same @param documentation to every function, so roxygen2 provides @inheritParams, which allows you to inherit argument documentation from another function, possibly even in another package.

stringr uses @inheritParams extensively because most functions have string and pattern arguments. The detailed and definitive documentation belongs to str_detect():

```
#' @param string Input vector. Either a character vector, or something
#'    coercible to one.
#' @param pattern Pattern to look for.
#'
#'    The default interpretation is a regular expression, as described in
#'    `vignette("regular-expressions")`. Use [regex()] for finer control of the
#'    matching behaviour.
#'
#'    Match a fixed string (i.e. by comparing only bytes), using
#'    [fixed()]. This is fast, but approximate. Generally,
#'    for matching human text, you'll want [coll()] which
#'    respects character matching rules for the specified locale.
#'
#'    Match character, word, line and sentence boundaries with
#'    [boundary()]. An empty pattern, "", is equivalent to
#'    `boundary("character")`.
```

Then the other stringr functions use @inheritParams str_detect to get this detailed documentation for string and pattern without having to duplicate that text.

@inheritParams inherits only docs for arguments that the function actually uses and that aren't already documented, so you can document some arguments locally and inherit others. str_match() uses this to inherit str_detect()'s standard documentation for the string argument, while providing its own specialized documentation for pattern:

```
#' @inheritParams str_detect
#' @param pattern Unlike other stringr functions, `str_match()` only supports
#'    regular expressions, as described `vignette("regular-expressions")`.
#'    The pattern should contain at least one capturing group.
```

Now that we've discussed default values and inheritance we can bring up one more dilemma. Sometimes there's tension between giving detailed information on an argument (acceptable values, default value, how the argument is used, etc.) and making the documentation amenable to reuse in other functions (which might differ in some specifics). This can motivate you to assess whether it's truly worth it for related functions to handle the same input in different ways or if standardization would be beneficial.

You can inherit documentation from a function in another package by using the standard :: notation, i.e., @inheritParams anotherpackage::function. This does introduce one small annoyance: now the documentation for your package is no longer self-contained and the version of anotherpackage can affect the generated docs. Beware of spurious diffs introduced by contributors who run devtools::document() with a different installed version of the inherited-from package.

Return Value

A function's output is as important as its inputs. Documenting the output is the job of the @returns[10] tag. Here the priority is to describe the overall "shape" of the output, i.e., what sort of object it is, and its dimensions (if that makes sense). For example, if your function returns a vector you might describe its type and length, or if your function returns a data frame you might describe the names and types of the columns and the expected number of rows.

The @returns documentation for functions in stringr is straightforward because almost all functions return some type of vector with the same length as one of the inputs. For example, here's how str_like() describes its output:

```
#' @returns A logical vector the same length as `string`.
```

A more complicated case is the joint documentation for str_locate() and str_locate_all().[11] str_locate() returns an integer matrix, and str_locate_all() returns a list of matrices, so the text needs to describe what determines the rows and columns:

---

10 For historical reasons, you can also use @return, but we now favor @returns because it reads more naturally.

11 We'll come back how to document multiple functions in one topic in "Multiple Functions in One Topic" on page 255.

```
#' @returns
#' * `str_locate()` returns an integer matrix with two columns and
#'    one row for each element of `string`. The first column, `start`,
#'    gives the position at the start of the match, and the second column, `end`,
#'    gives the position of the end.
#'
#'* `str_locate_all()` returns a list of integer matrices with the same
#'    length as `string`/`pattern`. The matrices have columns `start` and `end`
#'    as above, and one row for each match.
#' @seealso
#'    [str_extract()] for a convenient way of extracting matches,
#'    [stringi::stri_locate()] for the underlying implementation.
```

In other cases it can be easier to figure out what to highlight by thinking about the set of functions and how they differ. For example, most dplyr functions return a data frame, so just saying @returns A data frame is not very useful. Instead, we tried to identify exactly what makes each function different. We decided it makes sense to describe each function in terms of how it affects the rows, the columns, the groups, and the attributes. For example, this describes the return value of dplyr::filter():

```
#' @returns
#' An object of the same type as `.data`. The output has the following
#' properties:
#'
#' * Rows are a subset of the input, but appear in the same order.
#' * Columns are not modified.
#' * The number of groups may be reduced (if `.preserve` is not `TRUE`).
#' * Data frame attributes are preserved.
```

@returns is also a good place to describe any important warnings or errors that the user might see. For example, readr::read_csv() mentions what happens if there are any parsing problems:

```
#' @returns A [tibble()]. If there are parsing problems, a warning will
#'    alert you. You can retrieve the full details by calling [problems()]
#'    on your dataset.
```

Submitting to CRAN

For your initial CRAN submission, all functions must document their return value. While this may not be scrutinized in subsequent submissions, it's still a good practice. There's currently no way to check that you've documented the return value of every function (we're working on it (*https://oreil.ly/fc3GJ*)), which is why you'll notice some tidyverse functions lack output documentation. But we certainly aspire to provide this information across the board.

Examples

Describing what a function does is great, but *showing* how it works is even better. That's the role of the `@examples` tag, which uses executable R code to demonstrate what a function can do. Unlike other parts of the documentation where we've focused mainly on what you should write, here we'll briefly give some content advice and then focus mainly on the mechanics.

The main dilemma with examples is that you must jointly satisfy two requirements:

- Your example code should be readable and realistic. Examples are documentation that you provide for the benefit of the user, i.e., a real human, working interactively, trying to get their actual work done with your package.

- Your example code must run without error and with no side effects in many noninteractive contexts over which you have limited or no control, such as when CRAN runs R CMD check or when your package website is built via GitHub Actions.

It turns out that there is often tension between these goals and you'll need to find a way to make your examples as useful as you can for users, while also satisfying the requirements of CRAN (if that's your goal) or other automated infrastructure.

The mechanics of examples are complex because they must never error and they're executed in four different situations:

- Interactively using the `example()` function.

- During R CMD check on your computer, or another computer you control (e.g., in GitHub Actions).

- During R CMD check run by CRAN.

- When your pkgdown website is being built, often via GitHub Actions or similar.

After discussing what to put in your examples, we'll talk about keeping your examples self-contained, how to display errors if needed, handling dependencies, running examples conditionally, and alternatives to the `@examples` tag for including example code.

RStudio

When preparing *.R* scripts or *.Rmd/.qmd* reports, it's handy to use Ctrl/Cmd-Enter or the Run button to send a line of R code to the console for execution. Happily, you can use the same workflow for executing and developing the `@examples` in your roxygen comments. Remember to do `devtools::load_all()` often, to stay synced with the package source.

Contents

Use examples to first show the basic operation of the function, then to highlight any particularly important properties. For example, `str_detect()` starts by showing a few simple variations and then highlights a feature that's easy to miss—as well as passing a vector of strings and one pattern, you can also pass one string and vector of patterns:

```
#' @examples
#' fruit <- c("apple", "banana", "pear", "pineapple")
#' str_detect(fruit, "a")
#' str_detect(fruit, "^a")
#' str_detect(fruit, "a$")
#'
#' # Also vectorised over pattern
#' str_detect("aecfg", letters)
```

Try to stay focused on the most important features without getting into the weeds of every last edge case: if you make the examples too long, it becomes hard for the user to find the key application they're looking for. If you find yourself writing very long examples, it may be a sign that you should write a vignette instead.

There aren't any formal ways to break up your examples into sections, but you can use sectioning comments that use many hyphens to create a visual breakdown. Here's an example from `tidyr::chop()`:

```
#' @examples
#' # Chop --------------------------------------------------------------
#' df <- tibble(x = c(1, 1, 1, 2, 2, 3), y = 1:6, z = 6:1)
#' # Note that we get one row of output for each unique combination of
#' # non-chopped variables
#' df %>% chop(c(y, z))
#' # cf nest
#' df %>% nest(data = c(y, z))
#'
#' # Unchop ------------------------------------------------------------
#' df <- tibble(x = 1:4, y = list(integer(), 1L, 1:2, 1:3))
#' df %>% unchop(y)
#' df %>% unchop(y, keep_empty = TRUE)
```

Strive to keep the examples focused on the specific function that you're documenting. If you can make the point with a familiar built-in dataset, like mtcars, do so. If you find yourself needing to do a bunch of setup to create a dataset or object to use in the example, it may be a sign that you need to create a package dataset or even a helper function. See Chapter 7, "pkg_example() Path Helpers" on page 109, and "Create useful_things with a Helper Function" on page 220 for ideas. Making it easy to write (and read) examples will greatly improve the quality of your documentation.

Also, remember that examples are not tests. Examples should be focused on the authentic and typical usage you've designed for and that you want to encourage. The test suite is the more appropriate place to exhaustively exercise all of the arguments and to explore weird, pathological edge cases.

Leave the World as You Found It

Your examples should be self-contained. For example, this means:

- If you modify options(), reset them at the end of the example.
- If you create a file, create it somewhere in tempdir(), and make sure to delete it at the end of the example.
- Don't change the working directory.
- Don't write to the clipboard (unless a user is present to provide some form of consent).

This has a lot of overlap with our recommendations for tests (see "Self-Contained Tests" on page 205) and even for the R functions in your package (see "Respect the R Landscape" on page 91). However, due to the way that examples are run during R CMD check, the tools available for making examples self-contained are much more limited. Unfortunately, you can't use the withr package or even on.exit() to schedule clean up, like restoring options or deleting a file. Instead, you'll need to do it by hand. If you can avoid doing something that must then be undone, that is the best way to go and especially true for examples.

These constraints are often in tension with good documentation, if you're trying to document a function that somehow changes the state of the world. For example, you have to "show your work," i.e., all of your code, which means that your users will see all of the setup and teardown, even if it is not typical for authentic usage. If you're finding it hard to follow the rules, this might be another sign to switch to a vignette (see Chapter 17).

Submitting to CRAN

Many of these constraints are also mentioned in the CRAN repository policy (*https://oreil.ly/eQor-*), which you must adhere to when submitting to CRAN. Use the "find in page" feature to locate "malicious or anti-social" to see the details.

Additionally, you want your examples to send the user on a short walk, not a long hike. Examples need to execute relatively quickly so users can quickly see the results, it doesn't take ages to build your website, automated checks happen quickly, and it doesn't take up computing resources when submitting to CRAN.

Submitting to CRAN

All examples must run in under 10 minutes.

Errors

Your examples cannot throw any errors, so don't include flaky code that can fail for reasons beyond your control. In particular, it's best to avoid accessing websites, because R CMD check will fail whenever the website is down.

What can you do if you want to include code that causes an error for the purposes of teaching? There are two basic options:

- You can wrap the code in try() so that the error is shown but doesn't stop execution of the examples. For example, dplyr::bind_cols() uses try() to show you what happens if you attempt to column-bind two data frames with different numbers of rows:

    ```
    #' @examples
    #' ...
    #' # Row sizes must be compatible when column-binding
    #' try(bind_cols(tibble(x = 1:3), tibble(y = 1:2)))
    ```

- You can wrap the code in \dontrun{},[12] so it is never run by example(). The preceding example would look like this if you used \dontrun{} instead of try():

```
#' # Row sizes must be compatible when column-binding
#' \dontrun{
#' bind_cols(tibble(x = 1:3), tibble(y = 1:2)))
#' }
```

We generally recommend using try() so that the reader can see an example of the error in action.

Submitting to CRAN

For the initial CRAN submission of your package, all functions must have at least one example and the example code can't all be wrapped inside \dontrun{}. If the code can only be run under specific conditions, use the techniques in the next section to express those preconditions.

Dependencies and Conditional Execution

An additional source of errors in examples is the use of external dependencies: you can only use packages in your examples that your package formally depends on (i.e., that appear in Imports or Suggests). Furthermore, example code is run in the user's environment, not the package environment, so you'll have to either explicitly attach the dependency with library() or refer to each function with ::. For example, dbplyr is a dplyr extension package, so all of its examples start with library(dplyr):

```
#' @examples
#' library(dplyr)
#' df <- data.frame(x = 1, y = 2)
#'
#' df_sqlite <- tbl_lazy(df, con = simulate_sqlite())
#' df_sqlite %>% summarise(x = sd(x, na.rm = TRUE)) %>% show_query()
```

In the past, we recommended using code only from suggested packages inside a block like this:

```
#' @examples
#' if (requireNamespace("suggestedpackage", quietly = TRUE)) {
#'    # some example code
#' }
```

---

12 You used to be able to use \donttest{} for a similar purpose, but we no longer recommend it because CRAN sets a special flag that causes the code to be executed anyway.

We no longer believe that approach is a good idea, because:

- Our policy is to expect that suggested packages are installed when running R CMD check[13] and this informs what we do in examples, tests, and vignettes.

- The cost of putting example code inside { … } is high: you can no longer see intermediate results, such as when the examples are rendered in the package's website. The cost of a package not being installed is low: users can usually recognize the associated error and resolve it themselves, i.e., by installing the missing package.

In other cases, your example code may depend on something other than a package. For example, if your examples talk to a web API, you probably want to run them for an authenticated user, and you never want such code to run on CRAN. In this case, you really do need conditional execution. The entry-level solution is to express this explicitly:

```
#' @examples
#' if (some_condition()) {
#'   # some example code
#' }
```

The condition could be quite general, such as `interactive()`, or very specific, such as a custom predicate function provided by your package. But this use of `if()` still suffers from the downside highlighted previously, where the rendered examples don't clearly show what's going on inside the { … } block.

The `@examplesIf` tag is a great alternative to `@examples` in this case:

```
#' @examplesIf some_condition()
#' some_other_function()
#' some_more_functions()
```

This looks almost like the previous snippet, but it has several advantages:

- Users won't actually see the `if() { … }` machinery when they are reading your documentation from within R or on a pkgdown website. Users see only realistic code.

- The example code renders fully in pkgdown.

- The example code runs when it should and does not run when it should not.

- This doesn't run afoul of CRAN's prohibition of putting all your example code inside \dontrun{}.

---

13 This is certainly true for CRAN and is true in most other automated checking scenarios, such as our GitHub Actions workflows.

For example, googledrive (*https://oreil.ly/pvXrL*) uses `@examplesIf` in almost every function, guarded by `googledrive::drive_has_token()`. Here's how the examples for `googledrive::drive_publish()` begin:

```
#' @examplesIf drive_has_token()
#' # Create a file to publish
#' file <- drive_example_remote("chicken_sheet") %>%
#'    drive_cp()
#'
#' # Publish file
#' file <- drive_publish(file)
#' ...
```

The example code doesn't run on CRAN, because there's no token. It does run when the pkgdown site is built, because we can set up a token securely. And, if a normal user executes this code, they'll be prompted to sign in to Google, if they haven't already.

Intermixing Examples and Text

An alternative to examples is to use R Markdown code blocks elsewhere in your roxygen comments, either ```` ```R ```` if you just want to show some code, or ```` ```{r} ```` if you want the code to be run. These can be effective techniques, but there are downsides to each:

- The code in ```` ```R ```` blocks is never run; this means it's easy to accidentally introduce syntax errors or to forget to update it when your package changes.

- The code in ```` ```{r} ```` blocks is run every time you document the package. This has the nice advantage of including the output in the documentation (unlike examples), but the code can't take very long to run or your iterative documentation workflow will become quite painful.

Reusing Documentation

roxygen2 provides a number of features that allow you to reuse documentation across topics. They are documented in `vignette("reuse", package = "roxygen2")`, so here we'll focus on the three most important:

- Documenting multiple functions in one topic.

- Inheriting documentation from another topic.

- Using child documents to share prose between topics, or to share between documentation topics and vignettes.

Multiple Functions in One Topic

By default, each function gets its own documentation topic, but if two functions are very closely connected, you can combine the documentation for multiple functions into a single topic. For example, take `str_length()` and `str_width()`, which provide two different ways of computing the size of a string. As you can see from the description, both functions are documented together, because this makes it easier to see how they differ:

```
#' The length/width of a string
#'
#' @description
#' `str_length()` returns the number of codepoints in a string. These are
#' the individual elements (which are often, but not always letters) that
#' can be extracted with [str_sub()].
#'
#' `str_width()` returns how much space the string will occupy when printed
#' in a fixed width font (i.e. when printed in the console).
#'
#' ...
str_length <- function(string) {
  ...
}
```

To merge the two topics, `str_width()` uses `@rdname str_length` to add its documentation to an existing topic:

```
#' @rdname str_length
str_width <- function(string) {
  ...
}
```

This technique works best for functions that have a lot in common, i.e., similar return values and examples, in addition to similar arguments.

Inheriting Documentation

In other cases, functions in a package might share many related behaviors, but aren't closely enough connected that you want to document them together. We've discussed `@inheritParams` previously, but three variations allow you to inherit other things:

- `@inherit source_function` will inherit all supported components from `source_function()`.

- `@inheritSection source_function Section title` will inherit the single section with title "Section title" from `source_function()`.

- `@inheritDotParams` automatically generates parameter documentation for ... for the common case where you pass ... on to another function.

See *https://oreil.ly/4DeYr* for more details.

Child Documents

Finally, you can reuse the same *.Rmd* or *.md* document in the function documentation, *README.Rmd*, and vignettes by using R Markdown child documents. The syntax looks like this:

```
#' ```{r child = "man/rmd/filename.Rmd"}
#' ```
```

This is a feature we use very sparingly in the tidyverse, but one place we do use it is in dplyr, because a number of functions use the same syntax as `select()` and we want to provide all the info in one place:

```
#' # Overview of selection features
#'
#' ```{r, child = "man/rmd/overview.Rmd"}
#' ```
```

Then *man/rmd/overview.Rmd* contains the repeated markdown:

```
Tidyverse selections implement a dialect of R where operators make
it easy to select variables:

- `:` for selecting a range of consecutive variables.
- `!` for taking the complement of a set of variables.
- `&` and `|` for selecting the intersection or the union of two
  sets of variables.
- `c()` for combining selections.

...
```

If the *.Rmd* file contains roxygen (markdown-style) links to other help topics, then some care is needed. See *https://oreil.ly/YmsIh* for details.

Help Topic for the Package

This chapter focuses on function documentation, but remember you can document other things, as detailed in `vignette("rd-other", package = "roxygen2")`. In particular, you can create a help topic for the package itself by documenting the special sentinel `"_PACKAGE"`. The resulting *.Rd* file automatically pulls in information parsed from the *DESCRIPTION*, including title, description, list of authors, and useful URLs. This help topic appears alongside all your other topics and can also be accessed with `package?pkgname`, e.g., `package?usethis`, or even just `?usethis`.

We recommend calling `usethis::use_package_doc()` to set up this package-level documentation in a dummy file *R/{pkgname}-package.R*, whose contents will look something like:

```
#' @keywords internal
"_PACKAGE"
```

The *R/{pkgname}-package.R* file is the main reason we wanted to mention use_pack
age_doc() and package-level documentation here. It turns out there are a few other
package-wide housekeeping tasks for which this file is a very natural home. For
example, it's a sensible, central location for import directives, i.e., for importing
individual functions from your dependencies or even entire namespaces. In "In
Code Below R/" on page 158, we recommend importing specific functions via
usethis::use_import_from() and this function is designed to write the associated
roxygen tags into the *R/{pkgname}-package.R* file created by use_package_doc(). So,
putting it all together, this is a minimal example of how the *R/{pkgname}-package.R*
file might look:

```
#' @keywords internal
"_PACKAGE"

# The following block is used by usethis to automatically manage
# roxygen namespace tags. Modify with care!
## usethis namespace: start
#' @importFrom glue glue_collapse
## usethis namespace: end
NULL
```

## CHAPTER 17

Vignettes

A vignette is a long-form guide to your package. Function documentation is great if you know the name of the function you need, but it's useless otherwise. In contrast, a vignette can be framed around a target problem that your package is designed to solve. The vignette format is perfect for showing a workflow that solves that particular problem, start to finish. Vignettes afford you different opportunities than help topics: you have much more control over the integration of code and prose, and it's a better setting for showing how multiple functions work together.

Many existing packages have vignettes, and you can see all the vignettes associated with your installed packages with `browseVignettes()`. To limit that to a particular package, you can specify the package's name like so: `browseVignettes("tidyr")`. You can read a specific vignette with the `vignette()` function, e.g., `vignette("rectangle", package = "tidyr")`. To see vignettes for a package that you haven't installed, look at the "Vignettes" listing on its CRAN page (*https://oreil.ly/l4Fca*).

However, we much prefer to discover and read vignettes from a package's website, which is the topic of Chapter 19.[1] Compare the vignette experience of tidyr's CRAN page to what it feels like to access tidyr's vignettes from its website (*https://oreil.ly/PZ6J8*). Note that pkgdown uses the term "article," which feels like the right vocabulary for package users. The technical distinction between a vignette (which ships with a package) and an article (which is available only on the website; see "Article Instead of Vignette" on page 268) is something the package developer needs to think about. A pkgdown website presents all of the documentation of a package in a cohesive, interlinked way that makes it more navigable and useful. This chapter is ostensibly

---

1 This obviously depends on the quality of one's internet connection, so we make an effort to recommend behaviors that are compatible with base R's tooling around installed vignettes.

about vignettes, but the way we do things is heavily influenced by how those vignettes fit into a pkgdown website.

In this book, we're going to use R Markdown to write our vignettes,[2] just as we did for function documentation in "Key Markdown Features" on page 239. If you're not already familiar with R Markdown you'll need to learn the basics elsewhere; a good place to start is *https://rmarkdown.rstudio.com*.

In general, we embrace a somewhat circumscribed vignette workflow, i.e., there are many things that base R allows for that we simply don't engage in. For example, we treat *inst/doc/*[3] in the same way as *man/* and *NAMESPACE*, i.e., as something semi-opaque that is managed by automated tooling and that we don't modify by hand. Base R's vignette system allows for various complicated maneuvers that we just try to avoid. In vignettes, more than anywhere else, the answer to "But how do I do X?" is often "Don't do X."

Workflow for Writing a Vignette

To create your first vignette, run:

```
usethis::use_vignette("my-vignette")
```

This does the following:

1. Creates a *vignettes/* directory.
2. Adds the necessary dependencies to *DESCRIPTION*, i.e., adds knitr to the `VignetteBuilder` field and adds both knitr and rmarkdown to `Suggests`.
3. Drafts a vignette, *vignettes/my-vignette.Rmd*.
4. Adds some patterns to *.gitignore* to ensure that files created as a side effect of previewing your vignettes are kept out of source control (we'll say more about this later).

This draft document has the key elements of an R Markdown vignette and leaves you in a position to add your content. You also call `use_vignette()` to create your second and all subsequent vignettes; it will just skip any setup that's already been done.

---

2 Sweave is the original system used for authoring vignettes (Sweave files usually have extension *.Rnw*). Similar to our advice about how to author function documentation (see Chapter 16), we think it makes more sense to use a markdown-based syntax for vignettes than a one-off, LaTeX-associated format. This choice also affects the form of rendered vignettes: Sweave vignettes render to PDF, whereas R Markdown vignettes render to HTML. We recommend converting Sweave vignettes to R Markdown.

3 The *inst/doc/* folder is where vignettes go once they're built, when R `CMD build` makes the package bundle.

Once you have the draft vignette, the workflow is straightforward:

1. Start adding prose and code chunks to the vignette. Use `devtools::load_all()` as needed and use your usual interactive workflow for developing the code chunks.

2. Render the entire vignette periodically.

   This requires some intention, because unlike tests, by default, a vignette is rendered using the currently installed version of your package, not with the current source package, thanks to the initial call to `library(yourpackage)`.

   One option is to properly install your current source package with `devtools::install()` or, in RStudio, Ctrl/Cmd-Shift-B. Then use your usual workflow for rendering an *.Rmd* file. For example, press Ctrl/Cmd-Shift-K or click Knit.

   Or you could properly install your package and request that vignettes be built, with `install(build_vignettes = TRUE)`, then use `browseVignettes()`.

   Another option is for you to use `devtools::build_rmd("vignettes/my-vignette.Rmd")` to render the vignette. This builds your vignette against a (temporarily installed) development version of your package.

   It's very easy to overlook this issue and be puzzled when your vignette preview doesn't seem to reflect recent developments in the package. Double-check that you're building against the current version!

3. Rinse and repeat until the vignette looks the way you want.

If you're regularly checking your entire package (see "check() and R CMD check" on page 60), which we strongly recommend, this will help to keep your vignettes in good working order. In particular, this will alert you if a vignette uses a package that's not a formal dependency. We will come back to these package-level workflow issues in "How Vignettes Are Built and Checked" on page 269.

Metadata

The first few lines of the vignette contain important metadata. The default template contains the following information:

```
---
title: "Vignette Title"
output: rmarkdown::html_vignette
vignette: >
  %\VignetteIndexEntry{Vignette Title}
  %\VignetteEngine{knitr::rmarkdown}
  %\VignetteEncoding{UTF-8}
---
```

This metadata is written in YAML (*https://yaml.org*), a format designed to be both human and computer readable. YAML frontmatter is a common feature of R Markdown files. The syntax is much like that of the *DESCRIPTION* file, where each line consists of a field name, a colon, then the value of the field. The one special YAML feature we're using here is >. It indicates that the following lines of text are plain text and shouldn't use any special YAML features.

The default vignette template uses these fields:

title

> This is the title that appears in the vignette. If you change it, make sure to make the same change to VignetteIndexEntry{}. They should be the same, but unfortunately that's not automatic.

output

> This specifies the output format. There are many options that are useful for regular reports (including HTML, PDF, slideshows, etc.), but rmark down::html_vignette has been specifically designed for this exact purpose. See ?rmarkdown::html_vignette for more details.

vignette

> This is a block of special metadata needed by R. Here, you can see the legacy of LaTeX vignettes: the metadata looks like LaTeX comments. The only entry you might need to modify is the \VignetteIndexEntry{}. This is how the vignette appears in the vignette index, and it should match the title. Leave the other two lines alone. They tell R to use knitr to process the file and that the file is encoded in UTF-8 (the only encoding you should ever use for a vignette).

We generally don't use these fields, but you will see them in other packages:

author

> We don't use this unless the vignette is written by someone not already credited as a package author.

date

> We think this usually does more harm than good, since it's not clear what the date is meant to convey. Is it the last time the vignette source was updated? In that case you'll have to manage it manually and it's easy to forget to update it. If you manage date programmatically with Sys.date(), the date reflects when the vignette was built, i.e., when the package bundle was created, which has nothing to do with when the vignette or package was last modified. We've decided it's best to omit the date.

The draft vignette also includes two R chunks. The first one configures our preferred way of displaying code output and looks like this:

```{r, include = FALSE}
knitr::opts_chunk$set(
  collapse = TRUE,
  comment = "#>"
)
```

The second chunk just attaches the package the vignette belongs to:

```{r setup}
library(yourpackage)
```

You might be tempted to (temporarily) replace this `library()` call with `load_all()`, but we advise that you don't. Instead, use the techniques given in "Workflow for Writing a Vignette" on page 260 to exercise your vignette code with the current source package.

Advice on Writing Vignettes

> If you're thinking without writing, you only think you're thinking.
>
> —Leslie Lamport

When writing a vignette, you're teaching someone how to use your package. You need to put yourself in the reader's shoes and adopt a "beginner's mind." This can be difficult because it's hard to forget all of the knowledge that you've already internalized. For this reason, we find in-person teaching to be a really useful way to get feedback. You're immediately confronted with what you've forgotten that only you know.

A useful side effect of this approach is that it helps you improve your code. It forces you to re-see the initial onboarding process and to appreciate the parts that are hard. Our experience is that explaining how code works often reveals some problems that need fixing.

In fact, a key part of the tidyverse package release process is writing a blog post: we now do that before submitting to CRAN, because of the number of times it's revealed some subtle problem that requires a fix. It's also fair to say that the tidyverse and its supporting packages would benefit from more "how-to" guides, so that's an area where we are constantly trying to improve.

Writing a vignette also makes a nice break from coding. Writing seems to use a different part of the brain from programming, so if you're sick of programming, try writing for a bit.

Here are some resources we've found helpful:

- Literally anything written by Kathy Sierra. She is not actively writing at the moment, but her content is mostly timeless and is full of advice about programming, teaching, and how to create valuable tools. See her original blog, Creating Passionate Users (*https://headrush.typepad.com*), or the site that came after, Serious Pony (*https://seriouspony.com/blog*).

- *Style: Lessons in Clarity and Grace* by Joseph M. Williams and Joseph Bizup (Pearson Longman). This book helps you understand the structure of writing so that you'll be better able to recognize and fix bad writing.

Diagrams

It's a great idea to include relevant visualizations in your vignette, but these files can be fairly large.

Submitting to CRAN

You'll need to watch the file size. If you include a lot of graphics, it's easy to create a very large file. Be on the lookout for a NOTE that complains about an overly large directory. You might need to take explicit measures, such as lowering the resolution, reducing the number of figures, or switching from a vignette to an article (see "Article Instead of Vignette" on page 268).

Links

There is no official way to link to help topics from vignettes or vice versa or from one vignette to another.

This is a concrete example of why we think pkgdown sites are a great way to present package documentation, because pkgdown makes it easy (literally zero effort, in many cases) to get these hyperlinked cross-references. This is documented in `vignette("linking", package = "pkgdown")`. If you're reading this book online, the inline call to `vignette()` in the previous sentence should be hyperlinked to the corresponding vignette in pkgdown,[4] using the same toolchain that will create automatic links in your pkgdown websites! We discussed this syntax previously in "Key Markdown Features" on page 239, in the context of function documentation.

---

4 And, for anyone else, executing this code in the R console will open the vignette, if the host package is installed.

Automatic links are generated for functions in the host package, namespace-qualified functions in another package, vignettes, and more. Here are the two most important examples of automatically linked text:

`` `some_function()` ``
> Autolinked to the documentation of `some_function()`, within the pkgdown site of its host package. Note the use of backticks and the trailing parentheses.

`` `vignette("fascinating-topic")` ``
> Autolinked to the "fascinating-topic" article within the pkgdown site of its host package. Note the use of backticks.

Filepaths

Sometimes it is necessary to refer to another file from a vignette. The best way to do this depends on the application:

A figure created by code evaluated in the vignette
> By default, in the *.Rmd* workflow that we recommend, this takes care of itself. Such figures are automatically embedded into the *.html* using data URIs. You don't need to do anything. Example: `vignette("extending-ggplot2", package = "ggplot2")` generates a few figures in evaluated code chunks.

An external file that could be useful to users or elsewhere in the package (not just in vignettes)
> Put such a file in *inst/* (see "Installed Files" on page 116), perhaps in *inst/extdata/* (see "Raw Data File" on page 106), and refer to it with `system.file()` or `fs::path_package()` (see "Filepaths" on page 107). Example from `vignette("sf2", package = "sf")`:
>
> ````
> ```{r}
> library(sf)
> fname <- system.file("shape/nc.shp", package="sf")
> fname
> nc <- st_read(fname)
> ```
> ````

An external file whose utility is limited to your vignettes
> Put it alongside the vignette source files in *vignettes/* and refer to it with a filepath that is relative to *vignettes/*.

Example: The source of `vignette("tidy-data", package = "tidyr")` is found at *vignettes/tidy-data.Rmd* and it includes a chunk that reads a file located at *vignettes/weather.csv* like so:

```{r}
weather <- as_tibble(read.csv("weather.csv", stringsAsFactors = FALSE))
weather
```

An external graphics file

Put it in *vignettes/*, refer to it with a filepath that is relative to *vignettes/* and use `knitr::include_graphics()` inside a code chunk. Example from `vignette("sheet-geometry", package = "readxl")`:

```{r out.width = '70%', echo = FALSE}
knitr::include_graphics("img/geometry.png")
```

How Many Vignettes?

For simpler packages, one vignette is often sufficient. If your package is named "somepackage," call this vignette `somepackage.Rmd`. This takes advantage of a pkgdown convention, where the vignette that's named after the package gets an automatic "Get started" link in the top navigation bar.

More complicated packages probably need more than one vignette. It can be helpful to think of vignettes like chapters of a book—they should be self-contained but still link together into a cohesive whole.

Scientific Publication

Vignettes can also be useful if you want to explain the details of your package. For example, if you have implemented a complex statistical algorithm, you might want to describe all the details in a vignette so that users of your package can understand what's going on under the hood and be confident that you've implemented the algorithm correctly. In this case, you might also consider submitting your vignette to the *Journal of Statistical Software* (*http://jstatsoft.org*) or *The R Journal* (*http://journal.r-project.org*). Both journals are electronic only and peer-reviewed. Comments from reviewers can be very helpful for improving your package and vignette.

If you just want to provide something very lightweight so folks can easily cite your package, consider the *Journal of Open Source Software* (*https://joss.theoj.org*). This journal has a particularly speedy submission and review process, and it is where we published "Welcome to the Tidyverse," (*https://oreil.ly/I8ogw*) a paper we wrote so that folks could have a single paper to cite and all the tidyverse authors would get some academic credit.

Special Considerations for Vignette Code

A recurring theme is that the R code inside a package needs to be written differently from the code in your analysis scripts and reports. This is true for your functions (see "Understand When Code Is Executed" on page 87), tests (see "High-Level Principles for Testing" on page 203), and examples (see "Examples" on page 248), and it's also true for vignettes. In terms of what you can and cannot do, vignettes are fairly similar to examples, although some of the mechanics differ.

Any package used in a vignette must be a formal dependency, i.e., it must be listed in Imports or Suggests in *DESCRIPTION*. Similar to our stance in tests (see "In Test Code" on page 162), our policy is to write vignettes under the assumption that suggested packages will be installed in any context where the vignette is being built (see "In Examples and Vignettes" on page 163). We generally use suggested packages unconditionally in vignettes. But, as with tests, if a package is particularly hard to install, we might make an exception and take extra measures to guard its use.

There are many other reasons why it might not be possible to evaluate all of the code in a vignette in certain contexts, such as on CRAN's machines or in CI/CD. These include all the usual suspects: lack of authentication credentials, long-running code, or code that is vulnerable to intermittent failure.

The main method for controlling evaluation in an *.Rmd* document is the eval code chunk option, which can be TRUE (the default) or FALSE. Importantly, the value of eval can be the result of evaluating an expression. Here are some relevant examples:

- eval = requireNamespace("somedependency")
- eval = !identical(Sys.getenv("SOME_THING_YOU_NEED"), "")
- eval = file.exists("credentials-you-need")

The eval option can be set for an individual chunk, but in a vignette it's likely that you'll want to evaluate most or all of the chunks or practically none of them. In the latter case, you'll want to use knitr::opts_chunk$set(eval = FALSE) in an early, hidden chunk to make eval = FALSE the default for the remainder of the vignette. You can still override with eval = TRUE in individual chunks.

In vignettes, we use the eval option in a similar way as @examplesIf in examples (see "Dependencies and Conditional Execution" on page 252). If the code can be run only under specific conditions, you must find a way to check for those preconditions programmatically at runtime and use the result to set the eval option.

Here are the first few chunks in a vignette from googlesheets4, which wraps the Google Sheets API. The vignette code can only be run if we are able to decrypt a token that allows us to authenticate with the API. That fact is recorded in `can_decrypt`, which is then set as the vignette-wide default for `eval`:

````
```{r setup, include = FALSE}
can_decrypt <- gargle:::secret_can_decrypt("googlesheets4")
knitr::opts_chunk$set(
  collapse = TRUE,
  comment = "#>",
  error = TRUE,
  eval = can_decrypt
)
```

```{r eval = !can_decrypt, echo = FALSE, comment = NA}
message("No token available. Code chunks will not be evaluated.")
```

```{r index-auth, include = FALSE}
googlesheets4:::gs4_auth_docs()
```

```{r}
library(googlesheets4)
```
````

Notice the second chunk uses `eval = !can_decrypt`, which prints an explanatory message for anyone who builds the vignette without the necessary credentials.

The preceding example shows a few more handy chunk options. Use `include = FALSE` for chunks that should be evaluated but not seen in the rendered vignette. The `echo` option controls whether code is printed, in addition to output. Finally, `error = TRUE` is what allows you to purposefully execute code that could throw an error. The error will appear in the vignette, just as it would for your user, but it won't prevent the execution of the rest of your vignette's code, nor will it cause `R CMD check` to fail. This is something that works much better in a vignette than in an example.

Many other options are described at *https://yihui.name/knitr/options*.

Article Instead of Vignette

There is one last technique, if you don't want any of your code to execute on CRAN. Instead of a vignette, you can create an article, which is a term used by pkgdown for a vignette-like *.Rmd* document that is not shipped with the package, but that appears only in the website. An article will be less accessible than a vignette, for certain users, such as those with limited internet access, because it is not present in the local installation. But that might be an acceptable compromise, for example, for a package that wraps a web API.

You can draft a new article with `usethis::use_article()`, which ensures the article will be `.Rbuildignored`. A great reason to use an article instead of a vignette is to show your package working in concert with other packages that you don't want to depend on formally. Another compelling use case is when an article really demands lots of graphics. This is problematic for a vignette, because the large size of the package causes problems with `R CMD check` (and, therefore, CRAN) and is also burdensome for everyone who installs it, especially those with limited internet.

How Vignettes Are Built and Checked

We close this chapter by returning to a few workflow issues we didn't cover in "Workflow for Writing a Vignette" on page 260: How do the *.Rmd* files get turned into the vignettes consumed by users of an installed package? What does `R CMD check` do with vignettes? What are the implications for maintaining your vignettes?

It can be helpful to appreciate the big difference between the workflow for function documentation and vignettes. The source of function documentation is stored in roxygen comments in *.R* files below *R/*. We use `devtools::document()` to generate *.Rd* files below *man/*. These *man/*.Rd* files are part of the source package. The official R machinery cares *only* about the *.Rd* files.

Vignettes are very different because the *.Rmd* source is considered part of the source package, and the official machinery (`R CMD build` and `check`) interacts with vignette source and built vignettes in many ways. The result is that the vignette workflow feels more constrained, since the official tooling basically treats vignettes somewhat like tests, instead of documentation.

R CMD build and Vignettes

First, it's important to realize that the *vignettes/*.Rmd* source files exist only when a package is in source (see "Source Package" on page 34) or bundled form (see "Bundled Package" on page 35). Vignettes are rendered when a source package is converted to a bundle via `R CMD build` or a convenience wrapper such as `devtools::build()`. The rendered products (*.html*) are placed in *inst/doc/*, along with their source (*.Rmd*) and extracted R code (*.R*; discussed in "R CMD check and Vignettes" on page 271). Finally, when a package binary is made (see "Binary Package" on page 38), the *inst/doc/* directory is promoted to a top-level *doc/* directory, as happens with everything below *inst/*.

The key takeaway is that it is awkward to keep rendered vignettes in a source package, and this has implications for the vignette development workflow. It is tempting to fight this (and many have tried), but based on years of experience and discussion, the devtools philosophy is to accept this reality.

Assuming that you don't try to keep built vignettes around persistently in your source package, here are our recommendations for various scenarios:

- Active, iterative work on your vignettes. Use your usual interactive *.Rmd* workflow (such as the ✎ Knit button) or devtools::build_rmd("vignettes/my-vignette.Rmd") to render a vignette to *.html* in the *vignettes/* directory. Regard the *.html* as a disposable preview. (If you initiate vignettes with use_vignette(), this .html will already be gitignored.)

- Make the current state of vignettes in a development version available to the world:

  — Offer a pkgdown website, preferably with automated "build and deploy," such as using GitHub Actions to deploy to GitHub Pages. Here are tidyr's vignettes in the development version (note the "dev" in the URL): *https://tidyr.tidyverse.org/dev/articles/index.html*.

  — Be aware that anyone who installs directly from GitHub will need to explicitly request vignettes, e.g., with devtools::install_github(dependencies = TRUE, build_vignettes = TRUE).

- Make the current state of vignettes in a development version available locally:

  — Install your package locally and request that vignettes be built and installed, e.g., with devtools::install(dependencies = TRUE, build_vignettes = TRUE).

- Prepare built vignettes for a CRAN submission. Don't try to do this by hand or in advance. Allow vignette (re-)building to happen as part of devtools::submit_cran() or devtools::release(), both of which build the package.

If you really do want to build vignettes in the official manner on an ad hoc basis, devtools::build_vignettes() will do this. But we've seen this lead to developer frustration, because it leaves the package in a peculiar form that is a mishmash of a source package and an unpacked package bundle. This nonstandard situation can then lead to even more confusion. For example, it's not clear how these not-actually-installed vignettes are meant to be accessed. Most developers should avoid using build_vignettes() and, instead, pick one of the approaches just outlined.

Prebuilt Vignettes (or Other Documentation)

We highly recommend treating *inst/doc/* as a strictly machine-writable directory for vignettes. We recommend that you do not take advantage of the fact that you can place arbitrary prebuilt documentation in *inst/doc/*. This opinion permeates the devtools ecosystem which, by default, cleans out *inst/doc/* during various development tasks, to combat the problem of stale documentation.

However, we acknowledge that there are exceptions to every rule. In some domains, it might be impractical to rebuild vignettes as often as our recommended workflow implies. Here are a few tips:

- You can prevent the cleaning of *inst/doc/* with `pkgbuild::build(clean_doc =)`. You can put `Config/build/clean-inst-doc: FALSE` in DESCRIPTION to prevent pkgbuild and rcmdcheck from cleaning *inst/doc/*.
- The rOpenSci tech note How to precompute package vignettes or pkgdown articles (*https://oreil.ly/Xkwjd*) describes a clever, lightweight technique for keeping a manually updated vignette in *vignettes/*.
- The R.rsp (*https://oreil.ly/4tdxw*) package offers explicit support for static vignettes.

R CMD check and Vignettes

We conclude with a discussion of how vignettes are treated by R CMD check. This official checker expects a package bundle created by R CMD build, as described previously. In the devtools workflow, we usually rely on **devtools::check()**, which automatically does this build step for us, before checking the package. R CMD check has various command-line options and also consults many environment variables. We're taking a maximalist approach here, i.e., we describe all the checks that *could* happen.

R CMD check does some static analysis of vignette code and scrutinizes the existence, size, and modification times of various vignette-related files. If your vignettes use packages that don't appear in *DESCRIPTION*, that is caught here. If files that should exist don't exist or vice versa, that is caught here. This should not happen if you use the standard vignette workflow outlined in this chapter and is usually the result of some experiment that you've done, intentionally or not.

The vignette code is then extracted into a *.R* file, using the "tangle" feature of the relevant vignette engine (knitr, in our case), and run. The code originating from chunks marked as `eval = FALSE` will be commented out in this file and, therefore, is not executed. Then the vignettes are rebuilt from source, using the "weave" feature of the vignette engine (knitr, for us). This executes all the vignette code yet again, except for chunks marked eval = FALSE.

Submitting to CRAN

CRAN's incoming and ongoing checks use R CMD check which, as described, exercises vignette code up to two times. Therefore, it is important to conditionally suppress the execution of code that is doomed to fail on CRAN.

However, it's important to note that the package bundle and binaries distributed by CRAN actually use the built vignettes included in your submission. Yes, CRAN will attempt to rebuild your vignettes regularly, but this is for quality control purposes. CRAN distributes the vignettes you built.

## CHAPTER 18

Other Markdown Files

In this chapter we highlight two files that are conventionally used to provide some package-level documentation. These two are important, because they are featured on both the CRAN landing page and the pkgdown site for a package:

- *README.md*, which describes what the package does (see "README" on page 273). The *README* plays an especially important role on GitHub or similar platforms.
- *NEWS.md*, which describes how the package has changed over time (see "NEWS" on page 277).

Even if your package is intended for a very limited audience and might not ever be released on CRAN, these files can be very useful. These two files don't have to be written in markdown, but they can be. In keeping with our practices for help topics and vignettes, it's our strong recommendation and it's what we describe here.

README

First, we'll talk about the role of the *README* file and we leave off the file extension, until we're ready to talk about mechanics.

The goal of the *README* is to answer the following questions about your package:

- Why should I use it?
- How do I use it?
- How do I get it?

The *README* file is a long-established convention in software, going back decades. Some of its traditional content is found elsewhere in an R package; for example, we use the *DESCRIPTION* file to document authorship and licensing.

When you write your *README*, try to put yourself in the shoes of someone who has come across your package and is trying to figure out if it solves a problem they have. If they decide that your package looks promising, the *README* should also show them how to install it and how to do one or two basic tasks. Here's a good template for *README*:

1. A paragraph that describes the high-level purpose of the package.

2. An example that shows how to use the package to solve a simple problem.

3. Installation instructions, giving code that can be copied and pasted into R.

4. An overview that describes the main components of the package. For more complex packages, this will point to vignettes for more details. This is also a good place to describe how your package fits into the ecosystem of its target domain.

README.Rmd and README.md

As mentioned previously, we prefer to write *README* in markdown, i.e., to have *README.md*. This will be rendered as HTML and displayed in several important contexts:

- The repository home page, if you maintain your package on GitHub (or a similar host).
  - *https://github.com/tidyverse/dplyr*
- On CRAN, if you release your package there.
  - *https://cran.r-project.org/web/packages/dplyr/index.html*
    Notice the hyperlinked "README" under "Materials."
- As the home page of your pkgdown site, if you have one.
  - *https://dplyr.tidyverse.org*

Given that it's best to include a couple of examples in *README.md*, ideally you would generate it with R Markdown. That is, it works well to have *README.Rmd* as the main source file, which you then render to *README.md*.

The easiest way to get started is to use usethis::use_readme_rmd().[1] This creates a template *README.Rmd* and adds it to *.Rbuildignore*, since only *README.md* should be included in the package bundle. The template looks like this:

```
---
output: github_document
---

<!-- README.md is generated from README.Rmd. Please edit that file -->

```{r, include = FALSE}
knitr::opts_chunk$set(
  collapse = TRUE,
  comment = "#>",
  fig.path = "man/figures/README-",
  out.width = "100%"
)
```


somepackage

<!-- badges: start -->

<!-- badges: end -->

The goal of somepackage is to ...

Installation

You can install the development version of somepackage from
[GitHub](https://github.com/) with:

``` r
# install.packages("devtools")
devtools::install_github("jane/somepackage")
```

Example

This is a basic example which shows you how to solve a common problem:

```{r example}
library(somepackage)
## basic example code
```
```

---

1 If it really doesn't make sense to include any executable code chunks, usethis::use_readme_md() is similar, except that it gives you a basic *README.md* file.

```
What is special about using `README.Rmd` instead of just `README.md`?
You can include R chunks like so:

```{r cars}
summary(cars)
```


You'll still need to render `README.Rmd` regularly, to keep `README.md`
up-to-date. `devtools::build_readme()` is handy for this.

You can also embed plots, for example:

```{r pressure, echo = FALSE}
plot(pressure)
```


In that case, don't forget to commit and push the resulting figure files,
so they display on GitHub and CRAN.
```

A few things to note about this starter *README.Rmd*:

- It renders to GitHub Flavored Markdown (*https://github.github.com/gfm*).
- It includes a comment to remind you to edit *README.Rmd*, not *README.md*.
- It sets up our recommended knitr options, including saving images to *man/figures/README*—which ensures that they're included in your built package. This is important so that your *README* works when it's displayed by CRAN.
- It sets up a place for future badges, such as results from automatic continuous integration checks (see "Continuous Integration" on page 298). Examples of functions that insert development badges:
  — `usethis::use_cran_badge()` reports the current version of your package on CRAN.
  — `usethis::use_coverage()` reports test coverage.
  — `use_github_actions()` and friends report the R CMD check status of your development package.
- It includes placeholders where you should provide code for package installation and for some basic usage.
- It reminds you of key facts about maintaining your *README*.

You'll need to remember to rerender *README.Rmd* periodically and, most especially, before release. The best function to use for this is `devtools::build_readme()`, because it is guaranteed to render *README.Rmd* against the current source code of your package.

The devtools ecosystem tries to help you keep *README.Rmd* up-to-date in two ways:

- If your package is also a Git repo, `use_readme_rmd()` automatically adds the following precommit hook:

```bash
#!/bin/bash
if [[ README.Rmd -nt README.md ]]; then
  echo "README.md is out of date; please re-knit README.Rmd"
  exit 1
fi
```

  This prevents a `git commit` if *README.Rmd* is more recently modified than *README.md*. If the hook is preventing a commit you really want to make, you can override it with `git commit --no-verify`. Note that Git commit hooks are not stored in the repository, so this hook needs to be added to any fresh clone. For example, you could rerun `usethis::use_readme_rmd()` and discard the changes to *README.Rmd*.

- The release checklist placed by `usethis::use_release_issue()` includes a reminder to call `devtools::build_readme()`.

NEWS

The *README* is aimed at new users, whereas the *NEWS* file is aimed at existing users: it should list all the changes in each release that a user might notice or want to learn more about. As with *README*, it's a well-established convention for open source software to have a *NEWS* file, sometimes called a changelog.

As with *README*, base R tooling does not *require* that *NEWS* be a markdown file, but it does allow for that and it's our strong preference. A *NEWS.md* file is pleasant to read on GitHub, on your pkgdown site, and is reachable from your package's CRAN landing page. We demonstrate this again with dplyr:

- *NEWS.md* in dplyr's GitHub repo:

  — *https://github.com/tidyverse/dplyr/blob/main/NEWS.md*

- On CRAN, if you release your package there:

  — *https://cran.r-project.org/web/packages/dplyr/index.html*

    Notice the hyperlinked "NEWS" under "Materials."

- On your package site, available as the "Changelog" from the "News" drop-down menu in the main navbar:

  — *https://dplyr.tidyverse.org/news/index.html*

You can use `usethis::use_news_md()` to initiate the *NEWS.md* file; many other lifecycle- and release-related functions in the devtools ecosystem will make appropriate changes to *NEWS.md* as your package evolves.

Here's a hypothetical *NEWS.md* file:

```
# foofy (development version)

* Better error message when grooving an invalid grobble (#206).

# foofy 1.0.0

## Major changes

* Can now work with all grooveable grobbles!

## Minor improvements and bug fixes

* Printing scrobbles no longer errors (@githubusername, #100).

* Wibbles are now 55% less jibbly (#200).
```

This example demonstrates some organizing principles for *NEWS.md*:

- Use a top-level heading for each version: e.g., `# somepackage 1.0.0`. The most recent version should go at the top. Typically the top-most entry in *NEWS.md* of your source package will read `# somepackage (development version)`.[2]
- Each change should be part of a bulleted list. If you have a lot of changes, you might want to break them up using subheadings—`## Major changes`, `## Bug fixes`, etc.

  We usually stick with a simple list until we're close to a release, at which point we organize into sections and refine the text. It's hard to know in advance exactly what sections you'll need. The release checklist placed by `usethis::use_release_issue()` includes a reminder to polish the *NEWS.md* file. In that phase, it can be helpful to remember that *NEWS.md* is a user-facing record of change, in contrast to, e.g., commit messages, which are developer-facing.

---

2 pkgdown supports a few other wording choices for these headings; see more at *https://pkgdown.r-lib.org/reference/build_news.html*.

- If an item is related to an issue in GitHub, include the issue number in parentheses, e.g., (#10). If an item is related to a pull request, include the pull request number and the author, e.g., (#101, @hadley). This helps an interested reader to find relevant context on GitHub and, in your pkgdown site, these issue and pull request numbers and usernames will be hyperlinks. We generally omit the username if the contributor is already recorded in *DESCRIPTION*.

The main challenge with *NEWS.md* is getting into the habit of noting any user-visible change when you make it. It's especially easy to forget this when accepting external contributions. Before release, it can be useful to use your version control tooling to compare the source of the release candidate to the previous release. This often surfaces missing *NEWS* items.

## CHAPTER 19

Website

At this point, we've discussed many ways to document your package:

- Function documentation or, more generally, help topics (see Chapter 16).
- Documentation of datasets (see "Documenting Datasets" on page 104).
- Vignettes (and articles) (see Chapter 17).
- README and NEWS (see Chapter 18).

Wouldn't it be divine if all of that somehow got bundled up together into a beautiful website for your package? The pkgdown package (*https://pkgdown.r-lib.org*) is meant to provide exactly this magic, and that is the topic of this chapter.

Initiate a Site

Assuming your package has a valid structure, pkgdown should be able to make a website for it. Obviously that website will be more substantial if your package has more of the documentation elements just listed. But something reasonable should happen for any valid R package.

 We hear that some folks put off "learning pkgdown," because they think it's going to be a lot of work. But then they eventually execute the two commands we show next and have a decent website in less than five minutes!

`usethis::use_pkgdown()` is a function you run once and it does the initial, minimal setup necessary to start using pkgdown:

```
usethis::use_pkgdown()

#> ✓ Setting active project to '/private/tmp/RtmpRf8Oqf/mypackage'
#> ✓ Adding '^_pkgdown\\.yml$', '^docs$', '^pkgdown$' to '.Rbuildignore'
#> ✓ Adding 'docs' to '.gitignore'
#> ✓ Writing '_pkgdown.yml'
#> • Edit '_pkgdown.yml'
#> ✓ Setting active project to '<no active project>'
```

Here's what `use_pkgdown()` does:

- Creates *_pkgdown.yml*, which is the main configuration file for pkgdown. In an interactive session, *_pkgdown.yml* will be opened for inspection and editing. But there's no immediate need to change or add anything here.

- Adds various patterns to *.Rbuildignore*, to keep pkgdown-specific files and directories from being included in your package bundle.

- Adds *docs*, the default destination for a rendered site, to *.gitignore*. This is harmless for those who don't use Git. For those who do, this opts you in to our recommended lifestyle, where the definitive source for your pkgdown site is built and deployed elsewhere (probably via GitHub Actions and Pages; more on this soon). This means the rendered website at *docs/* just serves as a local preview.

`pkgdown::build_site()` is a function you'll call repeatedly to rerender your site locally. In an extremely barebones package, you'll see something like this:

```
pkgdown::build_site()

#> ✓ Setting active project to '/private/tmp/RtmpRf8Oqf/mypackage'
#> -- Installing package into temporary library ----------------------
#> == Building pkgdown site ==========================================
#> Reading from: '/private/tmp/RtmpRf8Oqf/mypackage'
#> Writing to:   '/private/tmp/RtmpRf8Oqf/mypackage/docs'
#> -- Initialising site ----------------------------------------------
#> Copying '../../../../Users/jenny/Library/R/...link.svg' to 'link.svg'
#> Copying '../../../../Users/jenny/Library/R/...pkgdown.js' to 'pkgdown.js'
#> -- Building home --------------------------------------------------
#> Writing 'authors.html'
#> Writing '404.html'
#> -- Building function reference ------------------------------------
#> Writing 'reference/index.html'
#> Writing 'sitemap.xml'
#> -- Building search index ------------------------------------------
#> == DONE ===========================================================
#> ✓ Setting active project to '<no active project>'
```

In an interactive session your newly rendered site should appear in your default web browser.

RStudio

Another nice gesture to build your site is via Addins > pkgdown > Build pkgdown.

You can look in the local *docs/* directory to see the files that constitute your package's website. To manually browse the site, open *docs/index.html* in your preferred browser.

This is almost all you truly need to know about pkgdown. It's certainly a great start and, as your package and ambitions grow, the best place to learn more is the pkgdown-made website (*https://pkgdown.r-lib.org*) for the pkgdown package itself.

Deployment

Your next task is to deploy your pkgdown site somewhere on the web, so that your users can visit it. The path of least resistance looks like this:

- Use Git and host your package on GitHub. The reasons to do this go well beyond offering a package website, but this will be one of the major benefits to adopting Git and GitHub, if you're on the fence.

- Use GitHub Actions (GHA) to build your website, i.e., to run `pkgdown::build_site()`. GHA is a platform where you can configure certain actions to happen automatically when some event happens. We'll use it to rebuild your website every time you push to GitHub.

- Use GitHub Pages to serve your website, i.e., the files you see below `docs/` locally. GitHub Pages is a static website hosting service that creates a site from files found in a GitHub repo.

The advice to use GitHub Action and Pages are implemented for you in the function `usethis::use_pkgdown_github_pages()`. It's not an especially difficult task, but there are several steps, and it would be easy to miss or flub one. The output of `use_pkgdown_github_pages()` should look something like this:

```
usethis::use_pkgdown_github_pages()
#> ✓ Initializing empty, orphan 'gh-pages' branch in GitHub repo
#>   'jane/mypackage'
#> ✓ GitHub Pages is publishing from:
#> • URL: 'https://jane.github.io/mypackage/'
#> • Branch: 'gh-pages'
#> • Path: '/'
#> ✓ Creating '.github/'
#> ✓ Adding '^\\.github$' to '.Rbuildignore'
#> ✓ Adding '*.html' to '.github/.gitignore'
#> ✓ Creating '.github/workflows/'
#> ✓ Saving 'r-lib/actions/examples/pkgdown.yaml@v2' to
```

```
#>   '.github/workflows/pkgdown.yaml'
#> • Learn more at <https://github.com/r-lib/actions/blob/v2/examples/README.md>.
#> ✔ Recording 'https://jane.github.io/mypackage/' as site's url in
#>   '_pkgdown.yml'
#> ✔ Adding 'https://jane.github.io/mypackage/' to URL field in DESCRIPTION
#> ✔ Setting 'https:/jane.github.io/mypackage/' as homepage of GitHub repo
#>   'jane/mypackage'
```

Like use_pkgdown(), this is a function you basically call once, when setting up a new site. In fact, the first thing it does is to call use_pkgdown() (it's OK if you've already called use_pkgdown()), so we usually skip straight to use_pkgdown_github_pages() when setting up a new site.

Let's walk through what use_pkgdown_github_pages() actually does:

- Initializes an empty, "orphan" branch in your GitHub repo, named gh-pages (for "GitHub Pages"). The gh-pages branch will live only on GitHub (there's no reason to fetch it to your local computer) and it represents a separate, parallel universe from your actual package source. The only files tracked in gh-pages are those that constitute your package's website (the files that you see locally below *docs/*).

- Turns on GitHub Pages for your repo and tells it to serve a website from the files found in the gh-pages branch.

- Copies the configuration file for a GHA workflow that does pkgdown "build and deploy." The file shows up in your package as *.github/workflows/pkgdown.yaml*. If necessary, some related additions are made to *.gitignore* and *.Rbuildignore*.

- Adds the URL for your site as the home page for your GitHub repo.

- Adds the URL for your site to *DESCRIPTION* and *_pkgdown.yml*. The autolinking behavior we've touted elsewhere relies on your package listing its URL in these two places, so this is a high-value piece of configuration.

After successful execution of use_pkgdown_github_pages(), you should be able to visit your new site at the URL displayed in the previous output.[1] By default the URL has this general form: *https://USERNAME.github.io/REPONAME/*.

Now What?

For a typical package, you could stop here—after creating a basic pkgdown site and arranging for it to be rebuilt and deployed regularly—and people using (or considering using) your package would benefit greatly. Everything beyond this point is a "nice to have."

---

1 Sometimes there's a small delay, so give it up to a couple of minutes to deploy.

Overall, we recommend `vignette("pkgdown", package = "pkgdown")` as a good place to start, if you think you want to go beyond the basic defaults.

In the following sections, we highlight a few areas that are connected to other topics in the book or customizations that are particularly rewarding.

Logo

It's fun to have a package logo! In the R community, we have a strong tradition of hex stickers, so it can be nice to join in with a hex logo of your own. Keen R user Amelia McNamara made herself a dress out of custom hex logo fabric (*https://oreil.ly/FevdZ*), and useR! 2018 featured a spectacular hex photo wall (*https://oreil.ly/J2q4_*).

Here are some resources to guide your logo efforts:

- The convention is to orient the logo with a vertex at the top and bottom, with flat vertical sides, as shown in Figure 19-1.

- If you think you might print stickers, make sure to comply with the de facto standard for sticker size. hexb.in (*http://hexb.in/sticker.html*) is a reliable source for the dimensions and also provides a list of potential vendors for printed stickers Figure 19-1.

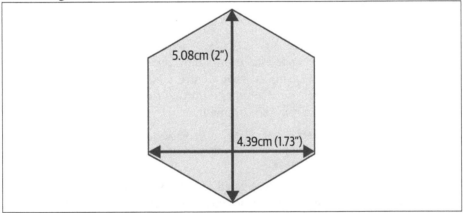

Figure 19-1. Standard dimensions of a hex sticker

- The hexSticker package (*https://oreil.ly/1WuiT*) helps you make your logo from within the comfort of R.

Once you have your logo, the usethis::use_logo() function places an appropriately scaled copy of the image file at *man/figures/logo.png* and provides a copy-paste-able markdown snippet to include your logo in your *README*. pkgdown will also discover a logo placed in the standard location and incorporate it into your site.

Reference Index

pkgdown creates a function reference in *reference/* that includes one page for each *.Rd* help topic in *man/*. This is one of the first pages you should admire in your new site. As you look around, there are a few things to contemplate, which we review in the following sections.

Rendered Examples

pkgdown executes all your examples (see "Examples" on page 248) and inserts the rendered results. We find this is a fantastic improvement over just showing the source code. This view of your examples can be eye-opening and often you'll notice things you want to add, omit, or change. If you're not satisfied with how your examples appear, this is a good time to review techniques for including code that is expected to error (see "Errors" on page 251) or that can be executed only under certain conditions (see "Dependencies and Conditional Execution" on page 252).

Linking

These help topics will be linked to from many locations within and, potentially, beyond your pkgdown site. This is what we are talking about in "Key Markdown Features" on page 239 when we recommend putting functions inside square brackets when mentioning them in a roxygen comment:

```
#' I am a big fan of [thisfunction()] in my package. I
#' also have something to say about [otherpkg::otherfunction()]
#' in somebody else's package.
```

On pkgdown sites, those square-bracketed functions become hyperlinks to the relevant pages in your pkgdown site. This is automatic within your package. But inbound links from *other* people's packages (and websites, etc.) require two things:[2]

---

2 Another prerequisite is that your package has been released on CRAN, because the auto-linking machinery has to look up the *DESCRIPTION* somewhere. It is possible to allow locally installed packages to link to each other, which is described in vignette("linking", package = "pkgdown").

- The URL field of your *DESCRIPTION* file must include the URL of your pkgdown site (preferably followed by the URL of your GitHub repo):

      URL: https://dplyr.tidyverse.org, https://github.com/tidyverse/dplyr

- Your *_pkgdown.yml* file must include the URL for your site:

      url: https://dplyr.tidyverse.org

devtools takes every chance it gets to do this sort of configuration for you. But if you elect to do things manually, this is something you might overlook. A general resource on auto-linking in pkgdown is `vignette("linking", package = "pkgdown")`.

Index Organization

By default, the reference index is just an alphabetically ordered list of functions. For packages with more than a handful of functions, it's often worthwhile to curate the index and organize the functions into groups. For example, dplyr uses this technique (*https://oreil.ly/f-eSp.*).

You achieve this by providing a `reference` field in *_pkgdown.yml*. Here's a redacted excerpt from dplyr's *_pkgdown.yml* file that gives you a sense of what's involved:

```
reference:
- title: Data frame verbs

- subtitle: Rows
  desc: >
    Verbs that principally operate on rows.
  contents:
  - arrange
  - distinct
  ...

- subtitle: Columns
  desc: >
    Verbs that principally operate on columns.
  contents:
  - glimpse
  - mutate
  ...

- title: Vector functions
  desc: >
    Unlike other dplyr functions, these functions work on individual vectors,
    not data frames.
  contents:
  - between
  - case_match
  ...
```

```
- title: Built in datasets
  contents:
  - band_members
  - starwars
  - storms
  ...

- title: Superseded
  desc: >
    Superseded functions have been replaced by new approaches that we believe
    to be superior, but we don't want to force you to change until you're
    ready, so the existing functions will stay around for several years.
  contents:
  - sample_frac
  - top_n
  ...
```

To learn more, see `?pkgdown::build_reference`.

Vignettes and Articles

Chapter 17 deals with vignettes, which are long-form guides for a package. They afford various opportunities beyond what's possible in function documentation. For example, you have much more control over the integration of prose and code and over the presentation of code itself; e.g., code can be executed but not seen, seen but not executed, and so on. It's much easier to create the reading experience that best prepares your users for authentic usage of your package.

A package's vignettes appear, in rendered form, in its website, in the Articles drop-down menu. "Vignette" feels like a technical term that we might not expect all R users to know, which is why pkgdown uses the term "articles" here. To be clear, the Articles menu lists your package's official vignettes (the ones that are included in your package bundle) and, optionally, other nonvignette articles (see "Article Instead of Vignette" on page 268), which are only available on the website.

Linking

Like function documentation, vignettes also can be the target of automatic inbound links from within your package and, potentially, beyond. We've talked about this elsewhere in the book. In "Key Markdown Features" on page 239, we introduced the idea of referring to a vignette with an inline call like `vignette("some-topic")`. The rationale behind this syntax is because the code can literally be copied, pasted, and executed for local vignette viewing. So it "works" in any context, even without automatic links. But, in contexts where the auto-linking machinery is available, it knows to look for this exact syntax and turn it into a hyperlink to the associated vignette, within a pkgdown site.

The need to specify the host package depends on the context:

`vignette("some-topic")`
> Use this form in your own roxygen comments, vignettes, and articles to refer to a vignette in your package. The host package is implied.

`vignette("some-topic", package = "somepackage")`
> Use this form to refer to a vignette in some other package. The host package must be explicit.

Note that this shorthand does *not* work for linking to nonvignette articles. Since the syntax leans so heavily on the `vignette()` function, it would be too confusing; i.e., evaluating the code in the console would fail because R won't be able to find such a vignette. Nonvignette articles must be linked like any other URL.

When you refer to a function in your package, in your vignettes and articles make sure to put it inside backticks and to include parentheses. Qualify functions from other packages with their namespace. Here's an example of prose in one of your own vignettes or articles:

```
I am a big fan of `thisfunction()` in my package. I also have something to
say about `otherpkg::otherfunction()` in somebody else's package.
```

Remember that automatic inbound links from other people's packages (and websites, etc.) require that your package advertises the URL of its website in *DESCRIPTION* and *_pkgdown.yaml*, as configured by `usethis::use_pkgdown_github_pages()` and as described in "Linking" on page 286.

Index Organization

As with the reference index, the default listing of the articles (broadly defined) in a package is alphabetical. But if your package has several articles, it can be worthwhile to provide additional organization. For example, you might feature the articles aimed at the typical user and tuck those meant for advanced users or developers behind "More articles ...". You can learn more about this in `?pkgdown::build_articles`.

NonVignette Articles

In general, Chapter 17 is our main source of advice on how to approach vignettes and that also includes some coverage of nonvignette articles ("Article Instead of Vignette" on page 268). Here we review some reasons to use a nonvignette article and give some examples.

An article is morally like a vignette (e.g., it tells a story that involves multiple functions and is written with R Markdown), except it does not ship with the package bundle. `usethis::use_article()` is the easiest way to create an article. The main reason to use an article is when you want to show code that is impossible or very painful to include in a vignette or official example. Possible root causes of this pain include:

- Use of a package you don't want to formally depend on. In vignettes and examples, it's forbidden to show your package working with a package that you don't list in *DESCRIPTION*, e.g., in `Imports` or `Suggests`.

  There is a detailed example of this in "Config/Needs/* Field" on page 166, featuring a readxl article that uses the tidyverse metapackage. The key idea is to list such a dependency in the `Config/Needs/website` field of *DESCRIPTION*. This keeps tidyverse out of readxl's dependencies, but ensures it's installed when the website is built.

- Code that requires authentication or access to specific assets, tools, or secrets that are not available on CRAN.

  The googledrive package (*https://oreil.ly/ZVtJk*) has no true vignettes, only non-vignette articles, because it's essentially impossible to demonstrate usage without authentication. It is possible to access secure environment variables on GitHub Actions, where the pkgdown site is built and deployed, but this is impossible to do on CRAN.

- Content that involves a lot of figures, which cause your package to bump up against CRAN's size constraints.

  The ggplot2 package presents several FAQs as articles (*https://oreil.ly/vaqWC*) for this reason.

Development Mode

Every pkgdown site has a so-called *development mode* (*https://oreil.ly/hNmQK*), which can be specified via the `development` field in *_pkgdown.yml*. If unspecified, the default is `mode: release`, which results in a single pkgdown site. Despite the name, this single site reflects the state of the current source package, which could be either a released state or a development state. The diagram below shows the evolution of a hypothetical package that is on CRAN and that has a pkgdown site in "release" mode:

```
...
 |
 V
Tweaks before release     v0.1.9000
 |
```

```
       V
Increment version number  v0.2.0      <-- install.packages() gets this
       |
       V
Increment version number  v0.2.9000
       |
       V
Improve error message     v0.2.9000  <-- site documents this
       |
       V
      ...
```

Users who install from CRAN get version 0.2.0. But the pkgdown site is built from the development version of the package.

This creates the possibility that users will read about some new feature on the website that is not present in the package version they have installed with install. packages(). We find that the simplicity of this setup outweighs the downsides, until a package has a broad user base, i.e., lots of users of varying levels of sophistication. It's probably safe to stay in "release" mode until you actually hear from a confused user.

Packages with a substantial user base should use "auto" development mode:

```
development:
  mode: auto
```

This directs pkgdown to generate a top-level site from the released version and to document the development version in a *dev/* subdirectory. We revisit the same hypothetical package as before, but we assume the pkdown site is in "auto" mode.

```
      ...
       |
       V
Tweaks before release     v0.1.9000
       |
       V
Increment version number  v0.2.0      <-- install.packages() gets this
       |                                   main site documents this
       V
Increment version number  v0.2.9000
       |
       V
Improve error message     v0.2.9000  <-- dev/ site documents this
       |
       V
      ...
```

All of the core tidyverse packages use "auto" mode. For example, consider the website of the readr package:

- *https://readr.tidyverse.org* documents the released version, i.e. what `install.packages("readr")` delivers.

- *https://readr.tidyverse.org/dev* documents the dev version, i.e., what `install_github("tidyverse/readr")` delivers.

Automatic development mode is recommended for packages with a broad user base, because it maximizes the chance that a user will read web-based documentation that reflects the package version that is locally installed.

Maintenance and Distribution

## CHAPTER 20

Software Development Practices

In this last part of the book, we zoom back out to consider development practices that can make you more productive and raise the quality of your work. Here we'll discuss the use of version control and continuous integration. In Chapter 21 we discuss how the nature of package maintenance varies over the lifecycle of a package.

You will notice that we recommend using certain tools:

An integrated development environment (IDE)
In "RStudio Projects" on page 52 we encouraged the use of the RStudio IDE for package development work. That's what we document, since it's what we use and devtools is developed to work especially well with RStudio. But even if it's not RStudio, we strongly recommend working with an IDE that has specific support for R and R package development.

Version control
We strongly recommend the use of formal version control and, at this point in time, Git is the obvious choice. We say that based on Git's general prevalence and, specifically, its popularity within the R package ecosystem. In "Git and GitHub" on page 296, we explain why we think version control is so important.

Hosted version control
We strongly recommend syncing your local Git repositories to a hosted service and, at this time, GitHub is "the" or at least "an" obvious choice. This is also covered in "Git and GitHub" on page 296.

Continuous integration and deployment, a.k.a. CI/CD (or even just CI)
This terminology comes from the general software engineering world and can sound somewhat grandiose or intimidating when applied to your personal R package. All this really means is that you set up specific package development tasks to happen automatically when you push new work to your hosted

repository. Typically you'll want to run R CMD check and to rebuild and deploy your package website. In "Continuous Integration" on page 298, we show how to do this with GitHub Actions.

You might think that these pro-style tools are overkill for someone who doesn't do software development for a living. While we don't recommend forcing yourself to do everything just listed on day one of your first "hello world" project, we actually do believe these tools are broadly applicable for R package development.

The main reason is that these tools make it so much easier to do the right thing, e.g., to experiment, document, test, check, and collaborate. By adopting a shared toolkit, part-time and newer package developers gain access to the same workflows used by experts. This requires a certain amount of faith and upfront investment, but we believe this pays off.

Git and GitHub

Git (*https://git-scm.com*) is a version control system that was originally built to coordinate the work of a global group of developers working on the Linux kernel. Git manages the evolution of a set of files—called a repository—in a highly structured way, and we recommend that every R package should also be a Git repository (and also, probably, an RStudio Project; see "RStudio Projects" on page 52).

A solo developer, working on a single computer, will benefit from adopting version control. But, for most of us, that benefit is not nearly large enough to make up for the pain of installing and using Git. In our opinion, for most folks, the pros of Git only outweigh the cons once you take the additional step of hooking your local repository up to a remote host like GitHub (*https://github.com*). The joint use of Git and GitHub offers many benefits that more than justify the learning curve.

Standard Practice

This recommendation is well aligned with the current, general practices in software development. Here are a few relevant facts from the 2022 Stack Overflow developer survey (*https://oreil.ly/TIdWx*), which is based on about 70K responses:

- 94% report using Git. The second-most used version control system was SVN, used by 5% of respondents.
- For personal projects, 87% of respondents report using GitHub, followed by GitLab (21%) and Bitbucket (11%). The ranking is the same albeit less skewed for professional work: GitHub still dominates with 56%, followed by GitLab (29%) and Bitbucket (18%).

We can even learn a bit about the habits of R package developers, based on the URLs found in the *DESCRIPTION* files of CRAN packages. As of March 2023, there are about 19K packages on CRAN, about 55% of which have a nonempty URL field (over 10K). Of those, 80% have a GitHub URL (over 8K), followed by GitLab (just over 1%) and Bitbucket (around 0.5%).

The prevalence of Git/GitHub, both within the R community and beyond, should help you feel confident that adoption will have tangible benefits. Furthermore, the sheer popularity of these tools means there are lots of resources available for learning how to use Git and GitHub and for getting unstuck.[1]

Two specific resources that address the intersection of Git/GitHub and the R world are the website Happy Git and GitHub for the useR (*https://oreil.ly/bl1Eo*) and the article "Excuse me, do you have a moment to talk about version control?"[2]

We conclude this section with a few examples of why Git/GitHub can be valuable specifically for R package development:

Communication with users
GitHub Issues are well-suited for taking bug reports and feature requests. Unlike email sent to the maintainer, these conversations are accessible to others and searchable.

Collaboration
GitHub pull requests are a very low-friction way for outside contributors to help fix bugs and add features.

Distribution
Functions like `devtools::install_github("r-lib/devtools")` and `pak::pak ("r-lib/devtools")` allow people to easily install the development version of your package, based on a source repository. More generally, anyone can install your package from any valid Git ref, such as a branch, specific SHA, pull request, or tag.

Website
GitHub Pages is one of the easiest ways to offer a website for your package (see "Deployment" on page 283).

Continuous integration
This is actually the topic of the next section, so read on for more.

---

1 We feature GitHub here, for hosted version control, because it's what we use and what has the best support in devtools. However, all the big-picture principles and even some details hold up for alternative platforms, such as GitLab and Bitbucket.

2 Jennifer Bryan, "Excuse Me, Do You Have a Moment to Talk about Version Control?" *The American Statistician* 72, no. 1 (2018):20–27. *https://doi.org/10.1080/00031305.2017.1399928*.

Continuous Integration

As we said in the introduction, continuous integration and deployment is commonly abbreviated as CI/CD or just CI. For R package development, what this means in practice is:

- You host your source package on a platform like GitHub. The key point is that the hosted repository provides the formal structure for integrating the work of multiple contributors. Sometimes multiple developers have permission to push (this is how tidyverse and r-lib packages are managed). In other cases, only the primary maintainer has push permission. In either model, external contributors can propose changes via a pull request.
- You configure one or more development tasks to execute automatically when certain events happen in the hosted repository, such as a push or a pull request. For example, for an R package, it's extremely valuable to configure an automatic run of R CMD check. This helps you discover breakage quickly, when it's easier to diagnose and fix, and is a tremendous help for evaluating whether to accept an external contribution.

Overall, the use of hosted version control and continuous integration can make development move more smoothly and quickly.

Even for a solo developer, having R CMD check run remotely, possibly on a couple of different operating systems, is a mighty weapon against the dreaded "works on my machine" problem. Especially for packages destined for CRAN, the use of CI decreases the chance of nasty surprises right before release.

GitHub Actions

The easiest way to start using CI is to host your package on GitHub and use its companion service, GitHub Actions (GHA). Then you can use various functions from usethis to configure so-called GHA workflows. usethis copies workflow configuration files from r-lib/actions (*https://github.com/r-lib/actions/#readme*), which is where the tidyverse team maintains GHA infrastructure useful to the R community.

R CMD check via GHA

If you use CI for only one thing, it should be to run R CMD check. If you call usethis::use_github_action() with no arguments, you can choose from a few of the most useful workflows. Here's what that menu looks like at the time of writing:

```
> use_github_action()
Which action do you want to add? (0 to exit)
(See <https://github.com/r-lib/actions/tree/v2/examples> for other options)

1: check-standard: Run `R CMD check` on Linux, macOS, and Windows
2: test-coverage: Compute test coverage and report to https://about.codecov.io
3: pr-commands: Add /document and /style commands for pull requests

Selection:
```

check-standard is highly recommended, especially for any package that is (or aspires to be) on CRAN. It runs R CMD check across a few combinations of operating system and R version. This increases your chances of quickly detecting code that relies on the idiosyncrasies of a specific platform, while it's still easy to make the code more portable.

After making that selection, you will see some messages along these lines:

```
#> ✓ Creating '.github/'
#> ✓ Adding '*.html' to '.github/.gitignore'
#> ✓ Creating '.github/workflows/'
#> ✓ Saving 'r-lib/actions/examples/check-standard.yaml@v2' to
#>    .github/workflows/R-CMD-check.yaml'
#> • Learn more at <https://github.com/r-lib/actions/blob/v2/examples/README.md>.
#> ✓ Adding R-CMD-check badge to 'README.md'
```

The key things that happen here are:

- A new GHA workflow file is written to *.github/workflows/R-CMD-check.yaml*. GHA workflows are specified via YAML files. The message reveals the source of the YAML and gives a link to learn more.

- Some helpful additions may be made to various "ignore" files.

- A badge reporting the R CMD check result is added to your *README*, if it has been created with usethis and has an identifiable badge "parking area." Otherwise, you'll be given some text you can copy and paste.

Commit these file changes and push to GitHub. If you visit the "Actions" section of your repository, you should see that a GHA workflow run has been launched. In due course, its success (or failure) will be reported there, in your *README* badge, and in your GitHub notifications (depending on your personal settings).

Congratulations! Your package will now benefit from even more regular checks.

Other Uses for GHA

As suggested by the interactive menu, usethis::use_github_action() gives you access to premade workflows other than R CMD check. In addition to the featured choices, you can use it to configure any of the example workflows in r-lib/

actions (*https://oreil.ly/Gjxh3*) by passing the workflow's name. For example, `use_github_action("test-coverage")` configures a workflow to track the test coverage of your package, as described in "Test Coverage" on page 202.

Since GHA allows you to run arbitrary code, you can use it for other things:

- Building your package's website and deploying the rendered site to GitHub Pages, as described in "Deployment" on page 283. See also `?usethis::use_pkgdown_github_pages()`.

- Republishing a book website every time you make a change to the source. (Like we do for this book!).

If the example workflows don't cover your exact use case, you can also develop your own workflow. Even in this case, the example workflows are often useful as inspiration. The `r-lib/actions` repository (*https://oreil.ly/mwPYg*) also contains important lower-level building blocks, such as actions to install R or to install all of the dependencies indicated in a *DESCRIPTION* file.

## CHAPTER 21

Lifecycle

This chapter is about managing the evolution of your package. The trickiest part of managing change is balancing the interests of various stakeholders:

- The maintainer(s), which includes you and possibly others, especially in the future.
- The existing users, which could be just you or a small group of colleagues or it could be tens or hundreds of thousands of people.
- The future users, which hopefully includes the existing users, but could potentially include many more people.

It's impossible to optimize for all of these folks, all of the time, all at once. So we'll describe how we think about various trade-offs. Even if your priorities differ from those of the tidyverse team, this chapter still should help you identify issues you want to consider.

Very few users complain when a package gains features or gets a bug fix. Instead, we're mostly going to talk about so-called breaking changes, such as removing a function or narrowing the acceptable inputs for a function. In "Backward Compatibility and Breaking Change" on page 307, we explore how to determine whether something is a breaking change or, more realistically, to gauge where it lies on a spectrum of "breakingness." Even though it can be painful, sometimes a breaking change is beneficial for the long-term health of a package (see "Pros and Cons of Breaking Change" on page 310).

Since change is inevitable, the kindest thing you can do for your users is to communicate clearly and help them adapt to change. Several practices work together to achieve this:

Package version number
> The main form of user-facing change is a package release. Be intentional about what sort of changes are included in, e.g., a patch release versus a major release (see "Package Version Number" on page 304, and "Major Versus Minor Versus Patch Release" on page 309).

Lifecycle stage
> Be explicit when a function or argument is regarded as experimental, superseded, or deprecated, as opposed to stable (the assumed default) (see "Lifecycle Stages and Supporting Tools" on page 312).

Deprecation process
> Enact change in a phased way, which makes it easier for users to adjust their code (see "Lifecycle Stages and Supporting Tools" on page 312).

Package Evolution

First we should establish a working definition of what it means for your package to change. Technically, you could say that the package has changed every time any file in its source changes. This level of pedantry isn't terribly useful, though. The smallest increment of change that's meaningful is probably a Git commit. This represents a specific state of the source package that can be talked about, installed from, compared to, subjected to R CMD check, reverted to, and so on. This level of granularity is really of interest only to developers. But the package states accessible via the Git history are genuinely useful for the maintainer, so if you needed any encouragement to be more intentional with your commits, let this be it.

The primary signal of meaningful change is to increment the package version number and release it, for some definition of release, such as releasing on CRAN (see Chapter 22). Recall that this important piece of metadata lives in the Version field of the *DESCRIPTION* file:

```
Package: usethis
Title: Automate Package and Project Setup
Version: 2.1.6
...
```

If you visit the CRAN landing page for usethis, you can access its history via Downloads > Old sources > usethis archive (*https://oreil.ly/ayCWm*). That links to a folder of package bundles (see "Bundled Package" on page 35), reflecting usethis's source for each version released on CRAN, presented in Table 21-1.

Table 21-1. Releases of the usethis package

Version	Date
1.0.0	2017-10-22 17:36:29 UTC
1.1.0	2017-11-17 22:52:07 UTC
1.2.0	2018-01-19 18:23:54 UTC
1.3.0	2018-02-24 21:53:51 UTC
1.4.0	2018-08-14 12:10:02 UTC
1.5.0	2019-04-07 10:50:44 UTC
1.5.1	2019-07-04 11:00:05 UTC
1.6.0	2020-04-09 04:50:02 UTC
1.6.1	2020-04-29 05:50:02 UTC
1.6.3	2020-09-17 17:00:03 UTC
2.0.0	2020-12-10 09:00:02 UTC
2.0.1	2021-02-10 10:40:06 UTC
2.1.0	2021-10-16 23:30:02 UTC
2.1.2	2021-10-25 07:30:02 UTC
2.1.3	2021-10-27 15:00:02 UTC
2.1.5	2021-12-09 23:00:02 UTC
2.1.6	2022-05-25 20:50:02 UTC

This is the type of package evolution we're going to address in this chapter. In "Package Version Number" on page 304, we'll delve into the world of software version numbers, which is a richer topic than you might expect. R also has some specific rules and tools around package version numbers. Finally, we'll explain the conventions we use for the version numbers of tidyverse packages (see "Tidyverse Package Version Conventions" on page 306).

But first, this is a good time to revisit a resource we first pointed out in "Source Package" on page 34, when introducing the different states of an R package. Recall that the (unofficial) cran organization on GitHub provides a read-only history of all CRAN packages. For example, you can get a different view of usethis's released versions at *https://github.com/cran/usethis*.

The archive provided by CRAN itself allows you to download older versions of use-this as *.tar.gz* files, which is useful if you truly want to get your hands on the source of an older version. However, if you just want to quickly check something about a version or compare two versions of usethis, the read-only GitHub mirror is much more useful. Each commit in this repo's history (*https://oreil.ly/6qKT3*) represents a CRAN release, which makes it easy to see exactly what changed. Furthermore, you can browse the state of all the package's source files at any specific version, such as usethis's initial release at version 1.0.0 (*https://oreil.ly/turrl*).[1]

This information is technically available from the repository (*https://oreil.ly/2Ujtu*) where usethis is actually developed. But you have to work much harder to zoom out to the level of CRAN releases, amid the clutter of the small incremental steps in which development actually unfolds. These three different views of usethis's evolution are all useful for different purposes:

https://cran.r-project.org/src/contrib/Archive/usethis
    The official CRAN package bundles.

https://github.com/cran/usethis/commits/HEAD
    The unofficial read-only CRAN mirror, obtained by unpacking CRAN's bundles.

https://github.com/r-lib/usethis/commits/HEAD
    The official development home for usethis.

Package Version Number

Formally, an R package version is a sequence of at least two integers separated by either . or -. For example, 1.0 and 0.9.1-10 are valid versions, but 1 and 1.0-devel are not. Base R offers the utils::package_version()[2] function to parse a package version string into a proper S3 class by the same name. This class makes it easier to do things like compare versions:

```
package_version(c("1.0", "0.9.1-10"))
#> [1] '1.0'      '0.9.1.10'
class(package_version("1.0"))
#> [1] "package_version" "numeric_version"

# these versions are not allowed for an R package
package_version("1")
#> Error: invalid version specification '1'
package_version("1.0-devel")
```

---

1 It's unusual for an initial release to be version 1.0.0, but remember that usethis was basically carved out of a very mature package (devtools).

2 We can call package_version() directly here, but in package code, you should use the utils::package_ver sion() form and list the utils package in Imports.

```
#> Error: invalid version specification '1.0-devel'

# comparing package versions
package_version("1.9") == package_version("1.9.0")
#> [1] TRUE
package_version("1.9") < package_version("1.9.2")
#> [1] TRUE
package_version(c("1.9", "1.9.2")) < package_version("1.10")
#> [1] TRUE TRUE
```

The previous examples make it clear that R considers version 1.9 to be equal to 1.9.0 and to be less than 1.9.2. And both 1.9 and 1.9.2 are less than 1.10, which you should think of as version "one point ten," not "one point one zero."

If you're skeptical that the package_version class is really necessary, check out this example:

```
"2.0" > "10.0"
#> [1] TRUE
package_version("2.0") > package_version("10.0")
#> [1] FALSE
```

The string 2.0 is considered to be greater than the string 10.0, because the character 2 comes after the character 1. By parsing version strings into proper package_version objects, we get the correct comparison, i.e., that version 2.0 is less than version 10.0.

R offers this support for working with package versions, because it's necessary, for example, to determine whether package dependencies are satisfied (see "Minimum Versions" on page 130). Under-the-hood, this tooling is used to enforce minimum versions recorded like this in *DESCRIPTION*:

```
Imports:
    dplyr (>= 1.0.0),
    tidyr (>= 1.1.0)
```

In your own code, if you need to determine which version of a package is installed, use utils::packageVersion():[3]

```
packageVersion("usethis")
#> [1] '2.2.0'
str(packageVersion("usethis"))
#> Classes 'package_version', 'numeric_version'  hidden list of 1
#>  $ : int [1:3] 2 2 0

packageVersion("usethis") > package_version("10.0")
#> [1] FALSE
```

---

3 As with package_version(), in package code, you should use the utils::packageVersion() form and list the utils package in Imports.

```
packageVersion("usethis") > "10.0"
#> [1] FALSE
```

The return value of `packageVersion()` has the `package_version` class and is therefore ready for comparison to other version numbers. Note the last example where we seem to be comparing a version number to a string. How can we get the correct result without explicitly converting `10.0` to a package version? It turns out this conversion is automatic as long as one of the comparators has the `package_version` class.

Tidyverse Package Version Conventions

R considers `0.9.1-10` to be a valid package version, but you'll never see a version number like that for a tidyverse package. Here is our recommended framework for managing the package version number:

- Always use `.` as the separator, never `-`.

- A released version number consists of three numbers, `<major>.<minor>.<patch>`. For version number `1.9.2`, `1` is the major number, `9` is the minor number, and `2` is the patch number. Never use versions like `1.0`. Always spell out the three components, `1.0.0`.

- An in-development package has a fourth component: the development version. This should start at 9000. The number 9000 is arbitrary, but provides a clear signal that there's something different about this version number. There are two reasons for this practice: First, the presence of a fourth component makes it easy to tell if you're dealing with a released or in-development version. Also, the use of the fourth place means that you're not limited to what the next released version will be. `0.0.1`, `0.1.0`, and `1.0.0` are all greater than `0.0.0.9000`.

  Increment the development version, e.g., from `9000` to `9001`, if you've added an important feature and you (or others) need to be able to detect or require the presence of this feature. For example, this can happen when two packages are developing in tandem. This is generally the only reason that we bother to increment the development version. This makes in-development versions special and, in some sense, degenerate. Since we don't increment the development component with each Git commit, the same package version number is associated with many different states of the package source, in between releases.

The preceding advice is inspired in part by Semantic Versioning (*https://semver.org*) and by the X.Org versioning schemes (*https://oreil.ly/i3yi4*). Read them if you'd like to understand more about the standards of versioning used by many open source projects. But we should underscore that our practices are inspired by these schemes and are somewhat less regimented. Finally, know that other maintainers follow different philosophies on how to manage the package version number.

Backward Compatibility and Breaking Change

The version number of your package is always increasing, but it's more than just an incrementing counter—the way the number changes with each release can convey information about the nature of the changes. The transition from 0.3.1 to 0.3.2, which is a patch release, has a very different vibe from the transition from 0.3.2 to 1.0.0, which is a major release. A package version number can also convey information about where the package is in its lifecycle. For example, the version 1.0.0 often signals that the public interface of a package is considered stable.

How do you decide which type of release to make, i.e., which component(s) of the version should you increment? A key concept is whether the associated changes are backward compatible, meaning that preexisting code will still "work" with the new version. We put "work" in quotes, because this designation is open to a certain amount of interpretation. A hardliner might take this to mean "the code works in exactly the same way, in all contexts, for all inputs." A more pragmatic interpretation is that "the code still works but could produce a different result in some edge cases." A change that is not backward compatible is often described as a *breaking* change. Here we're going to talk about how to assess whether a change is breaking. In "Pros and Cons of Breaking Change" on page 310 we'll talk about how to decide if a breaking change is worth it.

In practice, backward compatibility is not a clear-cut distinction. It is typical to assess the impact of a change from a few angles:

Degree of change in behavior
> The most extreme is to make something that used to be possible into an error, i.e., impossible.

How the changes fit into the design of the package
> A change to low-level infrastructure, such as a utility that gets called in all user-facing functions, is more fraught than a change that affects only one parameter of a single function.

How much existing usage is affected
> This is a combination of how many of your users will perceive the change and how many existing users there are to begin with.

Here are some concrete examples of breaking change:

- Removing a function
- Removing an argument
- Narrowing the set of valid inputs to a function

Conversely, these are usually not considered breaking:

- Adding a function. Caveat: there's a small chance this could introduce a conflict in user code.
- Adding an argument. Caveat: this could be breaking for some usage, e.g., if a user is relying on position-based argument matching. This also requires some care in a function that accepts . . .
- Increasing the set of valid inputs.
- Changing the text of a print method or error. Caveat: This can be breaking if other packages depend on yours in fragile ways, such as building logic or a test that relies on an error message from your package.
- Fixing a bug. Caveat: It really can happen that users write code that "depends" on a bug. Sometimes such code was flawed from the beginning, but the problem went undetected until you fixed your bug. Other times this surfaces code that uses your package in an unexpected way, i.e., it's not necessarily *wrong*, but neither is it *right*.

If reasoning about code was a reliable way to assess how it will work in real life, the world wouldn't have so much buggy software. The best way to gauge the consequences of a change in your package is to try it and see what happens. In addition to running your own tests, you can also run the tests of your reverse dependencies and see if your proposed change breaks anything. The tidyverse team has a fairly extensive set of tools for running so-called reverse dependency checks (see "Reverse Dependency Checks" on page 330), where we run R CMD check on all the packages that depend on ours. Sometimes we use this infrastructure to study the impact of a potential change, i.e., reverse dependency checks can be used to guide development, not only as a last-minute, prerelease check. This leads to yet another, deeply pragmatic definition of a breaking change:

> A change is breaking if it causes a CRAN package that was previously passing R CMD check to now fail AND the package's original usage and behavior is correct.

This is obviously a narrow and incomplete definition of breaking change, but at least it's relatively easy to get solid data.

Hopefully we've made the point that backward compatibility is not always a clearcut distinction. But hopefully we've also provided plenty of concrete criteria to consider when thinking about whether a change could break someone else's code.

Major Versus Minor Versus Patch Release

Recall that a version number will have one of these forms, if you're following the conventions described in "Tidyverse Package Version Conventions" on page 306:

```
<major>.<minor>.<patch>        # released version
<major>.<minor>.<patch>.<dev>  # in-development version
```

If the current package version is `0.8.1.9000`, here's our advice on how to pick the version number for the next release:

Increment `patch`, *e.g.,* `0.8.2` *for a patch release*
> You've fixed bugs, but you haven't added any significant new features and there are no breaking changes. For example, if we discover a show-stopping bug shortly after a release, we would make a quick patch release with the fix. Most releases will have a patch number of 0.

Increment `minor`, *e.g.,* `0.9.0`, *for a minor release*
> A minor release can include bug fixes, new features, and changes that are backward compatible.[4] This is the most common type of release. It's perfectly fine to have so many minor releases that you need to use two (or even three!) digits, e.g., `1.17.0`.

Increment `major`, *e.g.,* `1.0.0`, *for a major release*
> This is the most appropriate time to make changes that are not backward compatible and that are likely to affect many users. The `1.0.0` release has special significance and typically indicates that your package is feature complete with a stable API.

The trickiest decision you are likely to face is whether a change is "breaking" enough to deserve a major release. For example, if you make an API-incompatible change to a rarely used part of your code, it may not make sense to increase the major number. But if you fix a bug that many people depend on (it happens!), it will feel like a breaking change to those folks. It's conceivable that such a bug fix could merit a major release.

---

4 For some suitably pragmatic definition of "backward compatible."

We're mostly dwelling on breaking changes, but let's not forget that sometimes you also add exciting new features to your package. From a marketing perspective, you probably want to save these for a major release, because your users are more likely to learn about the new goodies from reading a blog post or *NEWS*.

Here are a few tidyverse blog posts that have accompanied different types of package releases:

- Major release: dplyr 1.0.0 (*https://oreil.ly/W2LMJ*), purrr 1.0.0 (*https://oreil.ly/QtwDi*), pkgdown 2.0.0 (*https://oreil.ly/4c3K4*), readr 2.0.0 (*https://oreil.ly/rEJpo*)
- Minor release: stringr 1.5.0 (*https://oreil.ly/A6faU*), ggplot2 3.4.0 (*https://oreil.ly/DgldB*)
- Patch release: These are usually not considered worthy of a blog post.

Package Version Mechanics

Your package should start with version number `0.0.0.9000`. `usethis::create_package()` starts with this version, by default.

From that point on, you can use `usethis::use_version()` to increment the package version. When called interactively, with no argument, it presents a helpful menu:

```
usethis::use_version()
#> Current version is 0.1.
#> What should the new version be? (0 to exit)
#>
#> 1: major --> 1.0
#> 2: minor --> 0.2
#> 3: patch --> 0.1.1
#> 4:   dev --> 0.1.0.9000
#>
#> Selection:
```

In addition to incrementing `Version` in *DESCRIPTION* (see Chapter 9), `use_version()` also adds a new heading in *NEWS.md* ("NEWS" on page 277).

Pros and Cons of Breaking Change

The big difference between major and minor releases is whether or not the code is backward compatible. In the general software world, the idea is that a major release signals to users that it may contain breaking changes and they should upgrade only when they have the capacity to deal with any issues that emerge.

Reality is a bit different in the R community, because of the way most users manage package installation. If we're being honest, most R users don't manage package versions in a very intentional way. Given the way update.packages() and install.packages() work, it's quite easy to upgrade a package to a new major version without really meaning to, especially for dependencies of the target package. This, in turn, can lead to unexpected exposure to breaking changes in code that previously worked. This unpleasantness has implications both for users and for maintainers.

If it's important to protect a data product against change in its R package dependencies, we recommend the use of a project-specific package library. In particular, we like to implement this approach using the renv package (*https://rstudio.github.io/renv/*). This supports a lifestyle where a user's default package library is managed in the usual, somewhat haphazard way. But any project that has a specific, higher requirement for reproducibility is managed with renv. This keeps package updates triggered by work in project A from breaking the code in project B and also helps with collaboration and deployment.

We suspect that package-specific libraries and tools like renv are currently underutilized in the R world. That is, lots of R users still use just one package library. Therefore, package maintainers still need to exercise considerable caution and care when they introduce breaking changes, regardless of what's happening with the version number. In the next section, we describe how tidyverse packages approach this, supported by tools in the lifecycle package.

As with dependencies (see "When Should You Take a Dependency?" on page 136), we find that extremism isn't a very productive stance. Extreme resistance to breaking change puts a significant drag on ongoing development and maintenance. Backward compatible code tends to be harder to work with because of the need to maintain multiple paths to support functionality from previous versions. The harder you strive to maintain backward compatibility, the harder it is to develop new features or fix old mistakes. This, in turn, can discourage adoption by new users and can make it harder to recruit new contributors. On the other hand, if you constantly make breaking changes, users will become very frustrated with your package and will decide they're better off without it. Find a happy medium. Be concerned about backward compatibility, but don't let it paralyze you.

The importance of backward compatibility is directly proportional to the number of people using your package: you are trading your time and pain for that of your users. There are good reasons to make backward incompatible changes. Once you've decided it's necessary, your main priority is to use a humane process that is respectful of your users.

Lifecycle Stages and Supporting Tools

The tidyverse team's approach to package evolution has become more structured and deliberate over the years. The associated tooling and documentation lives in the lifecycle package (*https://lifecycle.r-lib.org/index.html*). The approach relies on two major components:

- Lifecycle stages, which can be applied at different levels, i.e., to an individual argument or function or to an entire package. These stages are depicted in Figure 21-1.

- Conventions and functions to use when transitioning a function from one lifecycle stage to another. The deprecation process is the one that demands the most care.

We won't duplicate too much of the lifecycle documentation here. Instead, we highlight the general principles of lifecycle management and present specific examples of successful lifecycle "moves."

Lifecycle Stages and Badges

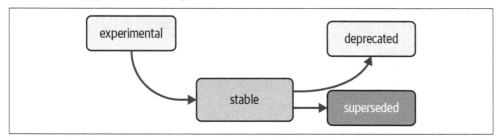

Figure 21-1. The four primary stages of the tidyverse lifecycle: stable, deprecated, superseded, and experimental

The four lifecycle stages are:

Stable
> This is the default stage and signals that users should feel comfortable relying on a function or package. Breaking changes should be rare and should happen gradually, giving users sufficient time and guidance to adapt their usage.

Experimental
> This is appropriate when a function is first introduced and the maintainer reserves the right to change it without much of a deprecation process. This is the implied stage for any package with a major version of 0, i.e., that hasn't had a 1.0.0 release yet.

Deprecated

This applies to functionality that is slated for removal. Initially, it still works, but it triggers a deprecation warning with information about preferred alternatives. After a suitable amount of time and with an appropriate version change, such functions are typically removed.

Superseded

This is a softer version of deprecated, where legacy functionality is preserved as if in a time capsule. Superseded functions receive only minimal maintenance, such as critical bug fixes.

You can get much more detail in `vignette("stages", package = "lifecycle")`.

The lifecycle stage is often communicated through a badge. If you'd like to use lifecycle badges, call `usethis::use_lifecycle()` to do some one-time setup:

```
usethis::use_lifecycle()
#> ✓ Adding 'lifecycle' to Imports field in DESCRIPTION
#> • Refer to functions with `lifecycle::fun()`
#> ✓ Adding '@importFrom lifecycle deprecated' to 'R/somepackage-package.R'
#> ✓ Writing 'NAMESPACE'
#> ✓ Creating 'man/figures/'
#> ✓ Copied SVG badges to 'man/figures/'
#> • Add badges in documentation topics by inserting one of:
#>    #' `r lifecycle::badge('experimental')`
#>    #' `r lifecycle::badge('superseded')`
#>    #' `r lifecycle::badge('deprecated')`
```

This leaves you in a position to use lifecycle badges in help topics and to use lifecycle functions, as described in the remainder of this section.

For a function, include the badge in its `@description` block. Here's how we indicate that `dplyr::top_n()` is superseded:

```
#' Select top (or bottom) n rows (by value)
#'
#' @description
#' `r lifecycle::badge("superseded")`
#' `top_n()` has been superseded in favour of ...
```

For a function argument, include the badge in the `@param` tag. Here's how the deprecation of `readr::write_file(path =)` is documented:

```
#' @param path `r lifecycle::badge("deprecated")` Use the `file` argument
#'   instead.
```

Call `usethis::use_lifecycle_badge()` if you want to use a badge in README to indicate the lifecycle of an entire package (see "README" on page 273).

If the lifecycle of a package is stable, it's not really necessary to use a badge, since that is the assumed default stage. Similarly, we typically use a badge for a function only if its stage differs from that of the associated package and likewise for an argument and the associated function.

Deprecating a Function

If you're going to remove or make significant changes to a function, it's usually best to do so in phases. Deprecation is a general term for the situation where something is explicitly discouraged, but it has not yet been removed. Various deprecation scenarios are explored in vignette("communicate", package = "lifecycle"); we're just going to cover the main idea here.

The lifecycle::deprecate_warn() function can be used inside a function to inform your user that they're using a deprecated feature and, ideally, to let them know about the preferred alternative. In this example, the plus3() function is being replaced by add3():

```
# new function
add3 <- function(x, y, z) {
  x + y + z
}

# old function
plus3 <- function(x, y, z) {
  lifecycle::deprecate_warn("1.0.0", "plus3()", "add3()")
  add3(x, y, z)
}

plus3(1, 2, 3)
#> Warning: `plus3()` was deprecated in somepackage 1.0.0.
#> i Please use `add3()` instead.
#> [1] 6
```

At this point, a user who calls plus3() sees a warning explaining that the function has a new name, but we go ahead and call add3() with their inputs. Preexisting code still "works." In some future major release, plus3() could be removed entirely.

lifecycle::deprecate_warn() and friends have a few features that are worth highlighting:

- The warning message is built up from inputs like when, what, with, and details, which gives deprecation warnings a predictable form across different functions, packages, and time. The intent is to reduce the cognitive load for users who may already be somewhat stressed.

- By default, a specific warning is issued once every 8 hours, in an effort to cause just the right amount of aggravation. The goal is to be just annoying enough to motivate the user to update their code before the function or argument goes away, but not so annoying that they fling their computer into the sea. Near the end of the deprecation process, the `always` argument can be set to TRUE to warn on every call.

- If you use `lifecycle::deprecate_soft()`, instead of `lifecycle::depre cate_warn()`, the warning is issued only if the person reading it is the one who can actually do something about it, i.e., update the offending code. If a user calls a deprecated function indirectly, i.e., because they are using a package that's using a deprecated function, by default that user doesn't get a warning. (But the maintainer of the guilty package will see these warnings in their test results.)

Here's a hypothetical schedule for removing a function `fun()`:

Package version `1.5.0`: `fun()` *exists*
> The lifecycle stage of the package is stable, as indicated by its post-`1.0.0` version number and, perhaps, a package-level badge. The lifecycle stage of `fun()` is also stable, by extension, since it hasn't been specifically marked as experimental.

Package version `1.6.0`
> The deprecation process of `fun()` begins. We insert `` `r lifecycle::badge ("deprecated")` `` in its `@description` to place a badge in its help topic. In the body of `fun()`, we add a call to `lifecycle::deprecate_warn()` to inform users about the situation. Otherwise, `fun()` still works as it always has.

Package version `1.7.0` *or* `2.0.0`
> `fun()` is removed. Whether this happens in a minor or major release will depend on the context, i.e., how widely used this package and function are.

If you're using base R only, the `.Deprecated()` and `.Defunct()` functions are the closest substitutes for `lifecycle::deprecate_warn()` and friends.

Deprecating an Argument

`lifecycle::deprecate_warn()` is also useful when deprecating an argument. In this case, it's also handy to use `lifecycle::deprecated()` as the default value for the deprecated argument. Here we continue an example from the preceding section, i.e., the switch from `path` to `file` in `readr::write_file()`:

```
write_file <- function(x,
                       file,
                       append = FALSE,
                       path = deprecated()) {
  if (is_present(path)) {
    lifecycle::deprecate_warn("1.4.0", "write_file(path)", "write_file(file)")
    file <- path
  }
```

```
      ...
    }
```

Here's what a user sees if they use the deprecated argument:

```
readr::write_file("hi", path = tempfile("lifecycle-demo-"))
#> Warning: The `path` argument of `write_file()` is deprecated as of readr
#> 1.4.0.
#> i Please use the `file` argument instead.
```

The use of deprecated() as the default accomplishes two things. First, if the user reads the documentation, this is a strong signal that an argument is deprecated. But deprecated() also has benefits for the package maintainer. Inside the affected function, you can use lifecycle::is_present() to determine if the user has specified the deprecated argument and proceed accordingly, as shown in the preceding code.

If you're using base R only, the missing() function has substantial overlap with lifecycle::is_present(), although it can be trickier to finesse issues around default values.

Deprecation Helpers

Sometimes a deprecation affects code in multiple places and it's clunky to inline the full logic everywhere. In this case, you might create an internal helper to centralize the deprecation logic.

This happened in googledrive, when we changed how to control the package's verbosity. The original design let the user specify this in every single function, via the verbose = TRUE/FALSE argument. Later, we decided it made more sense to use a global option to control verbosity at the package level. This is a case of (eventually) removing an argument, but it affects practically every single function in the package. Here's what a typical function looks like after starting the deprecation process:

```
drive_publish <- function(file, ..., verbose = deprecated()) {
  warn_for_verbose(verbose)
  # rest of the function ...
}
```

Note the use of verbose = deprecated(). Here's a slightly simplified version of warn_for_verbose():

```
warn_for_verbose <- function(verbose = TRUE,
                             env = rlang::caller_env(),
                             user_env = rlang::caller_env(2)) {
  # This function is not meant to be called directly, so don't worry about its
  # default of `verbose = TRUE`.
  # In authentic, indirect usage of this helper, this picks up on whether
  # `verbose` was present in the **user's** call to the calling function.
  if (!lifecycle::is_present(verbose) || isTRUE(verbose)) {
    return(invisible())
```

```
  }
  lifecycle::deprecate_warn(
    when = "2.0.0",
    what = I("The `verbose` argument"),
    details = c(
      "Set `options(googledrive_quiet = TRUE)` to suppress all googledrive
        messages.",
      "For finer control, use `local_drive_quiet()` or `with_drive_quiet()`.",
      "googledrive's `verbose` argument will be removed in the future."
    ),
    user_env = user_env
  )
  # only set the option during authentic, indirect usage
  if (!identical(env, global_env())) {
    local_drive_quiet(env = env)
  }
  invisible()
}
```

The user calls a function, such as drive_publish(), which then calls warn_for_ver
bose(). If the user leaves verbose unspecified or if they request verbose = TRUE
(default behavior), warn_for_verbose() does nothing. But if they explicitly ask for
verbose = FALSE, we throw a warning with advice on the preferred way to suppress
googledrive's messaging. We also go ahead and honor their wishes for the time being,
via the call to googledrive::local_drive_quiet(). In the next major release, the
verbose argument can be removed everywhere, and this helper can be deleted.

Dealing with Change in a Dependency

What if you want to use functionality in a new version of another package? Or
the less happy version: what if changes in another package are going to break your
package? There are a few possible scenarios, depending on whether the other package
has been released and the experience you want for your users. We'll start with the
simple, happier case of using features newly available in a dependency.

If the other package has already been released, you could bump the minimum version
you declare for it in *DESCRIPTION* and use the new functionality unconditionally.
This also means that users who update your package will be forced to update the
other package, which you should at least contemplate. Also note that this works only
for a dependency in Imports. While it's a good idea to record a minimum version for
a suggested package, it's not generally enforced the same as for Imports.

If you don't want to require your users to update this other package, you could make
your package work with both new and old versions. This means you'll check its
version at runtime and proceed accordingly. Here is a sketch of how that might look
in the context of an existing or new function:

```
your_existing_function <- function(..., cool_new_feature = FALSE) {
  if (isTRUE(cool_new_feature) && packageVersion("otherpkg") < "1.0.0") {
    message("otherpkg >= 1.0.0 is needed for cool_new_feature")
    cool_new_feature <- FALSE
  }
  # the rest of the function
}

your_new_function <- function(...) {
  if (packageVersion("otherpkg") < "1.0.0") {
    stop("otherpkg >= 1.0.0 needed for this function.")
  }
  # the rest of the function
}
```

Alternatively, this would also be a great place to use `rlang::is_installed()` and `rlang::check_installed()` with the `version` argument (see examples of usage in "In Code Below R/" on page 161).

This approach can also be adapted if you're responding to not-yet-released changes that are coming soon in one of your dependencies. It's helpful to have a version of your package that works both before and after the change. This allows you to release your package at any time, even before the other package. Sometimes you can refactor your code to make it work with either version of the other package, in which case you don't need to condition on the other package's version at all. But sometimes you might really need different code for the two versions. Consider this example:

```
your_function <- function(...) {
  if (packageVersion("otherpkg") >= "1.3.9000") {
    otherpkg::their_new_function()
  } else {
    otherpkg::their_old_function()
  }
  # the rest of the function
}
```

The hypothetical minimum version of `1.3.9000` suggests a case where the development version of otherpkg already has the change you're responding to, which is a new function in this case. Assuming `their_new_function()` doesn't exist in the latest release of otherpkg, you'll get a note from R CMD check stating that `their_new_function()` doesn't exist in otherpkg's namespace. If you're submitting such a version to CRAN, you can explain that you're doing this for the sake of backward and forward compatibility with otherpkg, and they are likely to be satisfied.

Superseding a Function

The last lifecycle stage that we'll talk about is superseded. This is appropriate when you feel like a function is no longer the preferred solution to a problem, but it has enough usage and history that you don't want to initiate the process of removing it.

Good examples of this are `tidyr::spread()` and `tidyr::gather()`. Those functions have been superseded by `tidyr::pivot_wider()` and `tidyr::pivot_longer()`. But some users still prefer the older functions, and it's likely that they've been used a lot in projects that are not under active development. Thus `spread()` and `gather()` are marked as superseded; they don't receive any new innovations, but they aren't at risk of removal.

A related phenomenon is when you want to change some aspect of a package, but you also want to give existing users a way to opt-in to the legacy behavior. The idea is to provide users a band-aid they can apply to get old code working quickly, until they have the bandwidth to do a more thorough update (which might not ever happen, in some cases). Here are some examples where legacy behavior was preserved for users who opt-in:

- In tidyr 1.0.0, the interface of `tidyr::nest()` and `tidyr::unnest()` changed. Most authentic usage can be translated to the new syntax, which tidyr does automatically, along with conveying the preferred modern syntax via a warning. But the old interface remains available via `tidyr:: nest_legacy()` and `tidyr::unn est_legacy()`, which were marked superseded upon creation.

- dplyr 1.1.0 takes advantage of a much faster algorithm for computing groups. But this speedier method also sorts the groups with respect to the C locale, whereas previously the system locale was used. The global option `dplyr.legacy_locale` allows a user to explicitly request the legacy behavior.[5]

- The tidyverse packages have been standardizing on a common approach to name repair, which is implemented in `vctrs::vec_as_names()`. The vctrs package also offers `vctrs::vec_as_names_legacy()`, which makes it easier to get names repaired with older strategies previously used in packages like tibble, tidyr, and readxl.

- readr 2.0.0 introduced a so-called second edition, marking the switch to a backend provided by the vroom package. Functions like `readr::with_edi tion(1, ...)` and `readr::local_edition(1)` make it easier for a user to request first-edition behavior for a specific bit of code or for a specific script.

---

5 You can learn more about the analysis leading up to this change in *https://oreil.ly/eJ_8w*.

## CHAPTER 22

Releasing to CRAN

We've been calling out CRAN-specific concerns throughout the book, on our journey through the various parts of a package, such as tests and examples. In this chapter, we focus on the actual process of releasing a package to CRAN, for the first time or as an update.

The most concrete expression of our release process is the checklist produced by `usethis::use_release_issue()`, which opens a GitHub issue containing a list of to-do's. This checklist is constantly evolving and is responsive to a few characteristics of the package, so don't be shocked if you see something a bit different than what we show here. The main concerns are fairly timeless, and we'll use this checklist to help structure this chapter.

But first: note that you will have deep regrets if you approach preparing your package for CRAN as a separate activity that you do *after* completing the planned development for a release. This advice is extremely relevant here:

> If it hurts, do it more often.[1]
>
> —Martin Fowler

In the current context, interpret this to mean that you should be running R CMD check regularly, preferably on multiple platforms, and promptly addressing any issues that surface. Recall that our preferred way to do this is via `devtools::check()` (see "check() and R CMD check" on page 60).

---

[1] Fowler's blog post "FrequencyReducesDifficulty" (*https://oreil.ly/rLiMh*) is a great read on this topic.

Why would you want to do something that is painful more often? Because it leads to less pain overall. First, solving five problems is more than five times as hard as solving one. It's demoralizing to be blocked by several errors and the potential for unsavory interactions between them makes each one harder to isolate and fix. Second, fast feedback tends to reduce your total number of mistakes. Once you learn some lesson the hard way, you are unlikely to make that same mistake dozens of times elsewhere in your package. Finally, practice makes perfect! With greater exposure, you will get better at interpreting and responding to problems that surface in R CMD check.

Another natural reaction is: why don't I eliminate this pain completely by not releasing my package on CRAN at all? For certain types of packages, this may be the right call. One such example is a personal package of helper functions. Another example is a package that supports a specific organization, as long as you also have a reasonable method of distributing that package to its users. The main reason to have your package on CRAN is to give your package greater reach. The vast majority of R users only install packages from CRAN, either due to personal or company policy or just from a lack of awareness about alternatives. CRAN provides discoverability, ease of installation, and a stamp of authenticity. The CRAN submission process can be frustrating, but it has many payoffs, and this chapter aims to make it as painless as possible.

The release process we describe here is best used as a preflight checklist that complements your ongoing efforts to keep your package passing R CMD check cleanly and CRAN-compliant. Two big realizations often come with a CRAN release process:

- If you've been turning a blind eye to WARNINGs and ERRORs from R CMD check, you really do have to study and eliminate those now. You should even eliminate as many NOTEs as possible.

- Even if your package passes R CMD check cleanly on your machine, it can be eye-opening when it leaves these cozy, familiar surroundings and is, instead, checked on a remote server, configured by someone else, running an entirely different operating system. This is why it is so valuable to use a continuous integration service like GitHub Actions (see "GitHub Actions" on page 298) to regularly check your package on macOS, Windows, and Linux.

These are the major steps in the release process:

1. Determine the release type, which dictates the version number.
2. If the package is already on CRAN: do due diligence on existing CRAN results. If this is a first release: confirm you are in compliance with CRAN policies.
3. Freshen up documentation files, such as *README.md* and *NEWS.md*.

4. Double check() that your package is passing cleanly on multiple operating systems and on the released and development version of R.

5. Perform reverse dependency checks, if other packages depend on yours.

6. Submit the package to CRAN and wait for acceptance.

7. Create a GitHub release and prepare for the next version by incrementing the version number.

8. Publicize the new version.

Decide the Release Type

When you call `use_release_issue()`, you'll be asked which type of release you intend to make:

```
> use_release_issue()
✓ Setting active project to '/Users/jenny/rrr/usethis'
Current version is 2.1.6.9000.
What should the release version be? (0 to exit)

1: major --> 3.0.0
2: minor --> 2.2.0
3: patch --> 2.1.7

Selection:
```

The immediate question feels quite mechanical: which component of the version number do you want to increment? But remember that we discussed the substantive differences in release types in "Major Versus Minor Versus Patch Release" on page 309.

In our workflow, this planned version number is recorded in the GitHub issue that holds the release checklist, but we don't actually increment the version in *DESCRIPTION* until later in the process (see "The Submission Process" on page 336). However, it's important to declare the release type up front, because the process (and, therefore, the checklist) looks different e.g., for a patch release versus a major release.

Initial CRAN Release: Special Considerations

Every new package receives a higher level of scrutiny from CRAN. In addition to the usual automated checks, new packages are also reviewed by a human, which inevitably introduces a certain amount of subjectivity and randomness. There are many packages on CRAN that would not be accepted in their current form, if submitted today as a completely new package. This isn't meant to discourage you. But you should be aware: just because you see some practice in an established package (or even in base R), that doesn't mean you can do the same in your new package.

Luckily, the community maintains lists of common "gotchas" for new packages. If your package is not yet on CRAN, the checklist begins with a special section that reflects this recent collective wisdom. Attending to these checklist items has dramatically improved our team's success rate for initial submissions.

First release:

- `usethis::use_news_md()`
- `usethis::use_cran_comments()`
- Update (aspirational) install instructions in *README*
- Proofread `Title:` and `Description:`
- Check that all exported functions have `@returns` and `@examples`
- Check that `Authors@R:` includes a copyright holder (role "cph")
- Check licensing of included files
- Review *https://github.com/DavisVaughan/extrachecks*

If you don't already have a *NEWS.md* file, you are encouraged to create one now with `usethis::use_news_md()`. You'll want this file eventually, and this anticipates the fact that the description of your eventual GitHub release (see "Celebrating Success" on page 338) is drawn from *NEWS.md*.

`usethis::use_cran_comments()` initiates a file to hold submission comments for your package. It's very barebones at first, e.g.:

```
## R CMD check results

0 errors | 0 warnings | 1 note

* This is a new release.
```

In subsequent releases, this file becomes less pointless; for example, it is where we report the results of reverse dependency checks. This is not a place to wax on with long explanations about your submission. In general, you should eliminate the need for such explanations, especially for an initial submission.

We highly recommend that your package have a *README* file (see "README" on page 273). If it does, this is a good time to check the installation instructions provided there. You may need to switch from instructions to install it from GitHub, in favor of installing from CRAN, in anticipation of your package's acceptance.

The `Title` and `Description` fields of *DESCRIPTION* are real hotspots for nitpicking during CRAN's human review. Carefully review the advice given in "Title and Description: What Does Your Package Do?" on page 125. Also check that `Authors@R` includes a copyright holder, indicated by the "cph" role. The two most common scenarios are that you add "cph" to your other roles (probably "cre" and "aut") or that you add your employer to `Authors@R:` with the "cph" and, perhaps, "fnd" role. (When you credit a funder via the "fnd" role, they are acknowledged in the footer of your pkgdown website.) This is also a good time to ensure that the maintainer's e-mail address is appropriate. This is the only way that CRAN can correspond with you. If there are problems and they can't get in touch with you, they will remove your package from CRAN. Make sure this email address is likely to be around for a while and that it's not heavily filtered.

Double-check that each of your exported functions documents its return value (with the `@returns` tag; see "Return Value" on page 246) and has an `@examples` section (see "Examples" on page 248). If you have examples that cannot be run on CRAN, you absolutely must use the techniques in "Dependencies and Conditional Execution" on page 252 to express the relevant preconditions properly. Do not take shortcuts, such as having no examples, commenting out your examples, or putting all of your examples inside `\dontrun{}`.

If you have embedded third-party code in your package, check that you are correctly abiding by and declaring its license (see "Code You Bundle" on page 178).

Finally, take advantage of any list of ad hoc checks that other package developers have recently experienced with CRAN. At the time of writing, *https://github.com/Davis Vaughan/extrachecks* is a good place to find such firsthand reports. Reading such a list and preemptively modifying your package often can make the difference between a smooth acceptance and a frustrating process requiring multiple attempts.

CRAN Policies

We alert you to specific CRAN policies throughout this book and, especially, through this chapter. However, this is something of a moving target, so it pays to make some effort to keep yourself informed about future changes to CRAN policy.

The official home of CRAN policy is *https://cran.r-project.org/web/packages/poli cies.html*. However, it's not very practical to read this document, e.g., once a week and simply hope that you'll notice any changes. The GitHub repository *https:// github.com/eddelbuettel/crp* monitors the CRAN Repository Policy by tracking the evolution of the underlying files in the source of the CRAN website. Therefore the commit history of that repository makes policy changes much easier to navigate. You also may want to follow the CRAN Policy Watch Mastodon account (*https://mas.to/ @CRANberriesFeed*), which toots whenever a change is detected.

The R-package-devel mailing list (*https://oreil.ly/NnCkR*) is another good resource for learning more about package development. You could subscribe to it to keep tabs on what other maintainers are talking about. Even if you don't subscribe, it can be useful to search this list when you're researching a specific topic.

Keeping Up with Change

Now we move into the main checklist items for a minor or major release of a package that is already on CRAN. Many of these items also appear in the checklist for a patch or initial release.

- Check current CRAN check results
- Check if any deprecation processes should be advanced, as described in Gradual deprecation (*https://oreil.ly/jgLTr*)
- Polish *NEWS* (*https://oreil.ly/HOnR-*)
- `urlchecker::url_check()`
- `devtools::build_readme()`

These first few items confirm that your package is keeping up with its surroundings and with itself. The first item, "Check current CRAN check results," will be a hyperlink to the CRAN check results for the version of the package that is currently on CRAN. If there are any WARNINGs or ERRORs or NOTEs there, you should investigate and determine what's going on. Occasionally there can be an intermittent hiccup at CRAN, but generally speaking, any result other than "OK" is something you should address with the release you are preparing. You may discover your package is in a dysfunctional state due to changes in base R, CRAN policies, CRAN tooling, or packages you depend on.

If you are in the process of deprecating a function or an argument, a minor or major release is a good time to consider moving that process along as described in "Lifecycle Stages and Supporting Tools" on page 312. This is also a good time to look at all the *NEWS* bullets that have accumulated since the last release ("Polish NEWS"). Even if you've been diligent about jotting down all the newsworthy changes, chances are these bullets will benefit from some reorganization and editing for consistency and clarity (see "NEWS" on page 277).

Another very important check is to run `urlchecker::url_check()`. CRAN's URL checks are described at *https://cran.r-project.org/web/packages/URL_checks.html* and are implemented by code that ships with R itself. However, these checks are not exposed in a very usable way. The urlchecker package was created to address this and exposes CRAN's URL-checking logic in the `url_check()` function. The main problems that surface tend to be URLs that don't work anymore or URLs that use redirection. Obviously, you should update or remove any URL that no longer exists. Redirection, however, is trickier. If the status code is "301 Moved Permanently," CRAN's view is that your package should use the redirected URL. The problem is that many folks don't follow RFC7231 (*https://oreil.ly/mYzc2*) to the letter and use this sort of redirect even when they have a different intent, i.e., their intent is to provide a stable, user-friendly URL that then redirects to something less user-friendly or more volatile. If a legitimate URL you want to use runs afoul of CRAN's checks, you'll have to choose between a couple of less-than-appealing options. You could try to explain the situation to CRAN, but this requires human review, and thus is not recommended. Or you can convert such URLs into nonhyperlinked, verbatim text. Note also that even though urlchecker is using the same *code* as CRAN, your local results may still differ from CRAN's, due to differences in other ambient conditions, such as environment variables and system capabilities.

If you have a *README.Rmd* file, you will also want to rebuild the static *README.md* file with the current version of your package. The best function to use for this is `dev tools::build_readme()` (*https://oreil.ly/pZssH*), because it is guaranteed to render *README.Rmd* against the current source code of your package.

Double R CMD Checking

Next come a couple of items related to `R CMD check`. Remember that this should not be the first time you've run `R CMD check` since the previous release! Hopefully, you are running `R CMD check` often during local development and are using a continuous integration service, like GitHub Actions. This is meant to be a last-minute, final reminder to double-check that all is still well:

devtools::check(remote = TRUE, manual = TRUE)
> This happens on your primary development machine, presumably with the current version of R, and with some extra checks that are usually turned off to make day-to-day development faster.

devtools::check_win_devel()

> This sends your package off to be checked with CRAN's win-builder service, against the latest development version of R (a.k.a. r-devel). You should receive an email within about 30 minutes with a link to the check results. It's a good idea to check your package with r-devel, because base R and R CMD check are constantly evolving. Checking with r-devel is required by CRAN policy, and it will be done as part of CRAN's incoming checks. There is no point in skipping this step and hoping for the best.

Note that the brevity of this list implicitly reflects that tidyverse packages are checked after every push via GitHub Actions, across multiple operating systems and versions of R (including the development version), and that most of the tidyverse team develops primarily on macOS. CRAN expects you to "make all reasonable efforts" to get your package working across all of the major R platforms, and packages that don't work on at least two will typically not be accepted.

The next subsection is optional reading with more details on all the platforms that CRAN cares about and how you can access them. If your ongoing checks are more limited than ours, you may want to make up for that with more extensive presubmission checks. You may also need this knowledge to troubleshoot a concrete problem that surfaces in CRAN's checks, either for an incoming submission or for a package that's already on CRAN.

When running R CMD check for a CRAN submission, you have to address any problems that show up:

- You must fix all ERRORs and WARNINGs. A package that contains any errors or warnings will not be accepted by CRAN.

- Eliminate as many NOTEs as possible. Each NOTE requires human oversight, which creates friction for both you and CRAN. If there are notes that you do not believe are important, it is almost always easier to fix them (even if the fix is a bit of a hack) than to persuade CRAN that they're OK. See our online-only guide to R CMD check (*https://r-pkgs.org/R-CMD-check.html*) for details on how to fix individual problems.

- If you can't eliminate a NOTE, list it in *cran-comments.md* and explain why you think it is spurious. We discuss this file further in "Update Comments for CRAN" on page 334.

  Note that there will always be one NOTE when you first submit your package. This reminds CRAN that this is a new submission and that they'll need to do some extra checks. You can't eliminate this NOTE, so just mention in *cran-comments.md* that this is your first submission.

CRAN Check Flavors and Related Services

CRAN runs `R CMD check` on all contributed packages upon submission and on a regular basis, on multiple platforms or what they call "flavors". You can see CRAN's current check flavors page (*https://oreil.ly/OdfzX*). There are various combinations of:

- Operating system and CPU: Windows, macOS (x86_64, arm64), Linux (various distributions)
- R version: r-devel, r-release, r-oldrel
- C, C++, FORTRAN compilers
- Locale, in the sense of the `LC_CTYPE` environment variable (this is about which human language is in use and character encoding)

CRAN's check flavors almost certainly include platforms other than your preferred development environment(s), so you will eventually need to make an explicit effort to check and perhaps troubleshoot your package on these other flavors.

It would be impractical for individual package developers to personally maintain all of these testing platforms. Instead, we turn to various community- and CRAN-maintained resources for this. Here is a selection, in order of how central they are to our current practices:

- GitHub Actions (GHA) is our primary means of testing packages on multiple flavors, as covered in "GitHub Actions" on page 298.
- R-hub builder (R-hub) is a service supported by the R Consortium where package developers can submit their package for checks that replicate various CRAN check flavors.

  You can use R-hub via a web interface (*https://builder.r-hub.io*) or, as we recommend, through the rhub R package (*https://r-hub.github.io/rhub/*).

  `rhub::check_for_cran()` is a good option for a typical CRAN package and is morally similar to the GHA workflow configured by `usethis::use_git hub_action("check-standard")`. However, unlike GHA, R-hub currently does not cover macOS, only Windows and Linux.

  rhub also helps you access some of the more exotic check flavors and offers specialized checks relevant to packages with compiled code, such as `rhub::check_with_sanitizers()`.

- macOS builder is a service maintained by the CRAN personnel who build the macOS binaries for CRAN packages. This is a relatively new addition to the list and checks packages with "the same setup and available packages as the CRAN M1 build machine."

You can submit your package using the web interface (*https://mac.r-project.org/macbuilder/submit.html*) or with `devtools::check_mac_release()`.

Reverse Dependency Checks

- `revdepcheck::revdep_check(num_workers = 4)`

This innocuous checklist item can actually represent a considerable amount of effort. At a high level, checking your **reverse dependencies** ("revdeps") breaks down into:

- Form a list of your reverse dependencies. These are CRAN packages that list your package in their `Depends`, `Imports`, `Suggests`, or `LinkingTo` fields.
- Run `R CMD` check on each one.
- Make sure you haven't broken someone else's package with the planned changes in your package.

Each of these steps can require considerable work and judgment. So, if you have no reverse dependencies, you should rejoice that you can skip this step. If you have only a couple of reverse dependencies, you can probably do this "by hand," i.e., download each package's source and run `R CMD` check.

Here we explain ways to do reverse dependency checks at scale, which is the problem we face. Some of the packages maintained by our team have thousands of reverse dependencies and even some of the lower-level packages have hundreds. We have to approach this in an automated fashion, and this section will be most useful to other maintainers in the same boat.

All of our reverse dependency tooling is concentrated in the revdepcheck package (*https://revdepcheck.r-lib.org/*). Note that, at least at the time of writing, the revdepcheck package is not on CRAN. You can install it from Github via `devtools::install_github("r-lib/revdepcheck")` or `pak::pak("r-lib/revdepcheck")`.

Do this when you're ready to do revdep checks for the first time:

```
usethis::use_revdep()
```

This does some one-time setup in your package's *.gitignore* and *.Rbuildignore* files. Revdep checking will create some rather large folders below `revdep/`, so you definitely want to configure these ignore files. You will also see this reminder to actually perform revdep checks like so, as the checklist item suggests:

```
revdepcheck::revdep_check(num_workers = 4)
```

This runs R CMD check on all of your reverse dependencies, with our recommendation to use four parallel workers to speed things along. The output looks something like this:

```
> revdepcheck::revdep_check(num_workers = 4)
— INIT ——————————————————————————— Computing revdeps —
— INSTALL ————————————————————————— 2 versions —
Installing CRAN version of cellranger
also installing the dependencies 'cli', 'glue', 'utf8', 'fansi', 'lifecycle',
'magrittr', 'pillar', 'pkgconfig', 'rlang', 'vctrs', 'rematch', 'tibble'

Installing DEV version of cellranger
Installing 13 packages: rlang, lifecycle, glue, cli, vctrs, utf8, fansi,
pkgconfig, pillar, magrittr, tibble, rematch2, rematch
— CHECK ——————————————————————————— 8 packages —
✓ AOV1R 0.1.0            — E: 0    | W: 0     | N: 0
✓ mschart 0.4.0          — E: 0    | W: 0     | N: 0
✓ googlesheets4 1.0.1    — E: 0    | W: 0     | N: 1
✓ readODS 1.8.0          — E: 0    | W: 0     | N: 0
✓ readxl 1.4.2           — E: 0    | W: 0     | N: 0
✓ readxlsb 0.1.6         — E: 0    | W: 0     | N: 0
✓ unpivotr 0.6.3         — E: 0    | W: 0     | N: 0
✓ tidyxl 1.0.8           — E: 0    | W: 0     | N: 0
OK: 8
BROKEN: 0
Total time: 6 min
— REPORT ——————————————————————————————
Writing summary to 'revdep/README.md'
Writing problems to 'revdep/problems.md'
Writing failures to 'revdep/failures.md'
Writing CRAN report to 'revdep/cran.md'
```

To minimize false positives, revdep_check() runs R CMD check twice per revdep: once with the released version of your package currently on CRAN and again with the local development version, i.e., with your release candidate. Why two checks? Because sometimes the revdep is already failing R CMD check and it would be incorrect to blame your planned release for the breakage. revdep_check() reports the packages that can't be checked and, most importantly, those where there are so-called "changes to the worse," i.e., where your release candidate is associated with new problems. Note also that revdep_check() always works with a temporary, self-contained package library, i.e., it won't modify your default user or system library.

tidyverse Team

We actually use a different function for our reverse dependency checks: `revdepcheck::cloud_check()`. This runs the checks in the cloud, massively in parallel, making it possible to run revdep checks for packages like testthat (with >10,000 revdeps) in just a few hours!

`cloud_check()` has been a gamechanger for us, allowing us to run revdep checks more often. For example, we even do this now when assessing the impact of a potential change to a package (see "Backward Compatibility and Breaking Change" on page 307), instead of only right before a release.

At the time of writing, `cloud_check()` is only available for package maintainers at Posit, but we hope to offer this service for the broader R community in the future.

In addition to some interactive messages, the revdep check results are written to the *revdep/* folder:

revdep/README.md
This is a high-level summary aimed at maintainers. The filename and markdown format are very intentional, in order to create a nice landing page for the *revdep* folder on GitHub.

revdep/problems.md
This lists the revdeps that appear to be broken by your release candidate.

revdep/failures.md
This lists the revdeps that could not be checked, usually because of an installation failure, either of the revdep itself or one of its dependencies.

revdep/cran.md
This is a high-level summary aimed at CRAN. You should copy and paste this into *cran-comments.md* (see "Update Comments for CRAN" on page 334).

checks.noindex, data.sqlite, library.noindex, and other files and folders
These are for revdepcheck's internal use and we won't discuss them further.

The easiest way to get a feel for these different files is to look around at the latest revdep results for some tidyverse packages, such as dplyr (*https://oreil.ly/g4p2A*) or tidyr (*https://oreil.ly/In5Gp*).

The revdep check results—local, cloud, or CRAN—are not perfect, because this is not a simple task. There are various reasons a result might be missing, incorrect, or contradictory in different runs:

False positives

Sometimes revdepcheck reports a package has been broken, but things are actually fine (or, at least, no worse than before). This most commonly happens because of flaky tests that fail randomly (see "Skip a Test" on page 228), such as HTTP requests. This can also happen because the instance runs out of disk space or other resources, so the first check using the CRAN version succeeds and the second check using the dev version fails. Sometimes it's obvious that the problem is not related to your package.

False negatives

Sometimes a package has been broken, but you don't detect that. For us, this usually happens when `cloud_check()` can't check a revdep because it can't be installed, typically because of a missing system requirement (e.g., Java). These are separately reported as "failed to test" but are still included in `problems.md`, because this could still be direct breakage caused by your package. For example, if you remove an exported function that's used by another package, installation will fail.

Generally these differences are less of a worry now that CRAN's own revdep checks are well automated, so new failures typically don't involve a human.

Revdeps and Breaking Changes

If the revdep check reveals breakages, you need to examine each failure and determine if it's:

- A false positive.
- A nonbreaking change, i.e., a failure caused by off-label usage of your package.
- A bug in your package that you need to fix.
- A deliberate breaking change.

If your update will break another package (regardless of why), you need to inform the maintainer, so they hear it first from you, rather than CRAN. The nicest way to do this is with a patch that updates their package to play nicely with yours, perhaps in the form of a pull request. This can be a decent amount of work and is certainly not feasible for all maintainers. But working through a few of these can be a good way to confront the pain that breaking change causes and to reconsider whether the benefits outweigh the costs. In most cases, a change that affects revdeps is likely to also break less visible code that lives outside of CRAN packages, such as scripts, reports, and Shiny apps.

If you decide to proceed, functions such as `revdepcheck::revdep_maintainers()` and `revdepcheck::revdep_email()` can help you notify revdep maintainers en masse. Make sure the email includes a link to documentation that describes the most common breaking changes and how to fix them. You should let the maintainers know when you plan to submit to CRAN (we recommend giving at least two weeks' notice), so they can submit their updated version before that. When your release date rolls around, re-run your checks to see how many problems have been resolved. Explain any remaining failures in *cran-comments.md* as demonstrated in "Update Comments for CRAN" on page 334. The two most common cases are that you are unable to check a package because you aren't able to install it locally or a legitimate change in the API that the maintainer hasn't addressed yet. As long as you have given sufficient advance notice, CRAN will accept your update, even if it breaks some other packages.

tidyverse Team

Lately the tidyverse team is trying to meet revdep maintainers more than halfway in terms of dealing with breaking changes. For example, in GitHub issue tidyverse/dplyr#6262 (*https://oreil.ly/ kdKSi*), the dplyr maintainers tracked hundreds of pull requests in the build-up to the release of dplyr v1.1.0. As the PRs are created, it's helpful to add links to those as well. As the revdep maintainers merge the PRs, they can be checked off as resolved. If some PRs are still in-flight when the announced submission date rolls around, the situation can be summarized in *cran-comments.md*, as was true in the case of dplyr v1.1.0 (*https://oreil.ly/I7K-F*).

Update Comments for CRAN

- Update *cran-comments.md*

We use the *cran-comments.md* file to record comments about a submission, mainly just the results from R CMD check and revdep checks. If you are making a specific change at CRAN's request, possibly under a deadline, that would also make sense to mention. We like to track this file in Git, so we can see how it changes over time. It should also be listed in *.Rbuildignore*, since it should not appear in your package bundle. When you're ready to submit, devtools::submit_cran() (see "The Submission Process" on page 336) incorporates the contents of *cran-comments.md* when it uploads your submission.

The target audience for these comments is the CRAN personnel, although there is no guarantee that they will read the comments (or when in the submission process they read them). For example, if your package breaks other packages, you will likely receive an automated email about that, even if you've explained it in the comments. Sometimes a human at CRAN then reads the comments, is satisfied, and accepts your package anyway, without further action from you. At other times, your package may be stuck in the queue until you copy *cran-comments.md* and paste it into an email exchange to move things along. In either case, it's worth keeping these comments in their own, version-controlled file.

Here is a fairly typical *cran-comments.md* from a recent release of forcats. Note that the R CMD check results are clean, i.e., there is nothing that needs to be explained or justified, and there is a concise summary of the revdep process:

```
## R CMD check results

0 errors | 0 warnings | 0 notes

## revdepcheck results

We checked 231 reverse dependencies (228 from CRAN + 3 from Bioconductor),
comparing R CMD check results across CRAN and dev versions of this package.

We saw 2 new problems:

* epikit
* stevemisc

Both maintainers were notified on Jan 12 (~2 week ago) and supplied with patches.

We failed to check 3 packages

* genekitr     (NA)
* OlinkAnalyze (NA)
* SCpubr       (NA)
```

This layout is designed to be easy to skim, and easy to match up to the R CMD check results seen by CRAN maintainers. It includes two sections:

Check results

We always state that there were no errors or warnings (and we make sure that's true!). Ideally we can also say there were no notes. But if not, any NOTEs are presented in a bulleted list. For each NOTE, we include the message from R CMD check and a brief description of why we think it's OK.

Here is how a NOTE is explained for the nycflights13 data package:

```
## R CMD check results

0 errors | 0 warnings | 1 note

* Checking installed package size:
    installed size is  6.9Mb
    sub-directories of 1Mb or more:
      data   6.9Mb

  This is a data package that will be rarely updated.
```

Reverse dependencies

> If there are revdeps, this is where we paste the contents of *revdep/cran.md* (see "Reverse Dependency Checks" on page 330). If there are no revdeps, we recommend that you keep this section, but say something like: "There are currently no downstream dependencies for this package."

The Submission Process

- `usethis::use_version('minor')` (or `'patch'` or `'major'`)
- `devtools::submit_cran()`
- Approve email

When you're truly ready to submit, it's time to actually bump the version number in *DESCRIPTION*. This checklist item will reflect the type of release declared at the start of this process (patch, minor, or major), in the initial call to `use_release_issue()`.

We recommend that you submit your package to CRAN by calling `devtools::submit_cran()`. This convenience function wraps up a few steps:

- Creates the package bundle (see "Bundled Package" on page 35) with `pkgbuild::build(manual = TRUE)`, which ultimately calls `R CMD build`.

- Posts the resulting **.tar.gz* file to CRAN's official submission form (*https://cran.r-project.org/submit.html*), populating your name and email from *DESCRIPTION* and your submission comments from *cran-comments.md*.

- Confirms that the submission was successful and reminds you to check your email for the confirmation link.

- Writes submission details to a local *CRAN-SUBMISSION* file, which records the package version, SHA, and time of submission. This information is used later by `usethis::use_github_release()` to create a GitHub release once your package has been accepted. *CRAN-SUBMISSION* will be added to *.Rbuildignore*. We generally do not gitignore this file, but neither do we commit it. It's an ephemeral note that exists during the interval between submission and (hopefully) acceptance.

After a successful upload, you should receive an email from CRAN within a few minutes. This email notifies you, as maintainer, of the submission and provides a confirmation link. Part of what this does is confirm that the maintainer's email address is correct. At the confirmation link, you are required to reconfirm that you've followed CRAN's policies and that you want to submit the package. If you fail to complete this step, your package is not actually submitted to CRAN!

Once your package enters CRAN's system it is automatically checked on Windows and Linux, probably against both the released and development versions of R. You will get another email with links to these check results, usually within a matter of hours. An initial submission (see "Initial CRAN Release: Special Considerations" on page 323) will receive additional scrutiny from CRAN personnel. The process is potentially fully automated when updating a package that is already on CRAN. If a package update passes its initial checks, CRAN will then run reverse dependency checks.

Failure Modes

There are at least three ways for your CRAN submission to fail:

- It does not pass R CMD check. This is an automated result.
- Human review finds the package to be in violation of CRAN policies. This applies mostly to initial submissions, but sometimes CRAN personnel decide to engage in ad hoc review of updates to existing packages that fail any automated checks.
- Reverse dependency checks suggest there are "changes to the worse." This is an automated result.

Failures are frustrating and the feedback may be curt and may feel downright insulting. Take comfort in the fact that this a widely shared experience across the R community. It happens to us on a regular basis. Don't rush to respond, especially if you are feeling defensive.

Wait until you are able to focus your attention on the technical issues that have been raised. Read any check results or emails carefully and investigate the findings. Unless you feel extremely strongly that discussion is merited, don't respond to the e-mail. Instead:

- Fix the identified problems and make recommended changes. Rerun dev tools::check() on any relevant platforms to make sure you didn't accidentally introduce any new problems.
- Increase the patch version of your package. Yes, this means that there might be gaps in your released version numbers. This is not a big deal.
- Add a "Resubmission" section at the top of *cran-comments.md*. This should clearly identify that the package is a resubmission, and list the changes that you made:

  ```
  ## Resubmission
  This is a resubmission. In this version I have:
  ```

```
* Converted the DESCRIPTION title to title case.

* More clearly identified the copyright holders in the DESCRIPTION
  and LICENSE files.
```

- If necessary, update the check results and revdep sections.

- Run `devtools::submit_cran()` to resubmit the package.

If your analysis indicates that the initial failure was a false positive, reply to CRAN's email with a concise explanation. For us, this scenario mostly comes up with respect to revdep checks. It's extremely rare for us to see failure for CRAN's initial R CMD check runs and, when it happens, it's often legitimate. On the other hand, for packages with a large number of revdeps, it's inevitable that a subset of these packages have some flaky tests or brittle examples. Therefore it's quite common to see revdep failures that have nothing to do with the proposed package update. In this case, it is appropriate to send a reply email to CRAN explaining why you think these are false positives.

Celebrating Success

Now we move into the happiest section of the checklist:

- Accepted 🎉
- `git push`
- `usethis::use_github_release()`
- `usethis::use_dev_version()`
- `git push`
- Finish blog post, share on social media, etc.
- Add link to blog post in pkgdown news menu

CRAN will notify you by email once your package is accepted. This is when we first push to GitHub with the new version number, i.e., we wait until it's certain that this version will actually be released on CRAN. Next we create a GitHub release corresponding to this CRAN release, using `usethis::use_github_release()`. A GitHub release is basically a glorified Git tag. The only aspect of GitHub releases that we regularly take advantage of is the release notes. `usethis::use_github_release()` creates release notes from the *NEWS* bullets relevant to the current release. Note that `usethis::use_github_release()` depends crucially on the *CRAN-SUBMISSION* file that was written by `devtools::submit_cran()`: that's how it knows which SHA to tag. After the successful creation of the GitHub release, `use_github_release()` deletes this temporary file.

Now we prepare for the next release by incrementing the version number yet again, this time to a development version using `usethis::use_dev_version()`. It makes sense to immediately push this state to GitHub so that, for example, any new branches or pull requests clearly have a development version as their base.

After the package has been accepted by CRAN, binaries are built for macOS and Windows. It will also be checked across the panel of CRAN check flavors. These processes unfold over a few days post-acceptance, and sometimes they uncover errors that weren't detected by the less comprehensive incoming checks. It's a good idea to visit your package's CRAN landing page a few days after release and just make sure that all still seems to be well. Figure 22-1 highlights where these results are linked from a CRAN landing page.

Figure 22-1. Link to CRAN check results

If there is a problem, prepare a patch release to address it and submit using the same process as before. If this means you are making a second submission less than a week after the previous, explain the situation in *cran-comments.md*. Getting a package established on CRAN can take a couple of rounds, although the guidance in this chapter is intended to maximize the chance of success on the first try. Future releases, initiated from your end, should be spaced at least one or two months apart, according to CRAN policy.

Once your package's binaries are built and it has passed checks across CRAN's flavors, it's time for the fun part: publicizing your package. This takes different forms, depending on the type of release. If this is your initial release (or, at least, the first release for which you really want to attract users), it's especially important to spread the word. No one will use your helpful new package if they don't know it exists. There are a number of places to announce your package, such as Twitter, Mastodon, LinkedIn, Slack communities, etc. Make sure to use any relevant tags, such as the #rstats hashtag. If you have a blog, it's a great idea to write a post about your release.

When introducing a package, the vibe should be fairly similar to writing your *README* or a "Get Started" vignette. Make sure to describe what the package does, so that people who haven't used it before can understand why they should even care. For existing packages, we tend to write blog posts for minor and major releases, but not for a patch release. In all cases, we find that these blog posts are most effective when they include lots of examples, i.e., "show, don't tell." For package updates, remember that the existence of a comprehensive *NEWS* file frees you from the need to list every last change in your blog post. Instead, you can focus on the most important changes and link to the full release notes, for those who want the gory details.

If you do blog about your package, it's good to capture this as yet another piece of documentation in your pkgdown website. A typical pkgdown site has a "News" item in the top navbar, linking to a "Changelog," which is built from *NEWS.md*. This drop-down menu is a common place to insert links to any blog posts about the package. You can accomplish this by having YAML like this in your *_pkgdown.yml* configuration file:

```
news:
  releases:
  - text: "Renaming the default branch (usethis >= 2.1.2)"
    href: https://www.tidyverse.org/blog/2021/10/renaming-default-branch/
  - text: "usethis 2.0.0"
    href: https://www.tidyverse.org/blog/2020/12/usethis-2-0-0/
  - text: "usethis 1.6.0"
    href: https://www.tidyverse.org/blog/2020/04/usethis-1-6-0/
```

Congratulations! You have released your first package to CRAN and made it to the end of the book!

Index

README.md, 274-277, 327
README.Rmd, 20-23, 274-277, 327
readxl::readxl_example(), 109
recursive dependencies, 137
redirected URLs in package, 327
reference index, website, 286-287
regexcite, 2
regexcite.Rproj, 5
regular expressions, 2, 16
release process, CRAN, 321-340
    celebrating success, 338-340
    double R CMD checking, 327-330
    failure modes, 337
    initial release considerations, 323-326
    keeping up with change, 326
    policies, 112, 228, 230, 251, 325
    release type decision, 323
    revdep checks, 330-336
    submission process, 336
    title and description for, 125
    update comments for CRAN, 334
relicensing, 177
rematch2 R package, 2
Remotes field, DESCRIPTION, 165
remotes package, 41
renv package, 311
.Renviron file, 45
repeat code, in testing, 210, 219
reproducibility, CRAN considerations for testing, 229
require(), 91, 151
requireNamespace(), 152, 163
return value, function documentation, 246
@returns, roxygen tag, 246, 325
reusing documentation, 254-256
revdepcheck package, 330-336
revdepcheck::cloud_check(), 332
revdepcheck::revdep_check, 330
revdepcheck::revdep_email(), 333
revdepcheck::revdep_maintainers(), 333
reverse dependencies, 209, 308, 330-336, 335, 337
rex R package, 2
.Rinstignore, 42
rJava R package, 137
rlang::check_installed(), 162, 318
rlang::is_installed(), 162, 318
.Rmd versus .Rd files and vignettes, 269
roles, in Authors@R field, 127

roxygen2 package, 12
    (see also function documentation)
    comments, blocks, tags, 237-239
    documentation workflow, 234-237
    documenting datasets, 104
    markdown features, 239-240
    and NAMESPACE file, 145
    and namespace tags, 145, 156-157, 159
    .Rd file generation, 234
roxygen2::roxygenise(), 235
.Rprofile, 29
.Rproj file, 54
.Rproj.user, 5
RStudio, xvi-xvii
    example code, 249
    function definition storage, 85
    installing package with, 41
    pkgdown build in, 283
    test runs in, 189
RStudio Desktop, 27
RStudio IDE, 27
RStudio Project, 4, 52-56
run time versus build time, 74-75

S

S3 object system, 169-171
S3method(), NAMESPACE directive, 144
s3_register(), 172
S4 object system, 145, 169
scripts
    and code execution, 87-91
    function name conventions, 83-85
    other users' use of, 91-96
    package versus script exercise, 63-65
    versus packages, 87-91
sd() and var(), 143
search path, dependencies, 146-151
search(), 146
secrets, testing, 227
self-contained tests, 205-208
self-sufficient tests, 203-205
setup files, testthat, 214
side effects, 76-78, 95-96
skipping a test, 61, 224-226, 228
skip_on_cran(), 228
snapshot tests, 195-198, 229
software development practices, 47, 295-300
    (see also workflows and tooling)
source package, 34, 35-37, 51

Visual Studio Code (VS Code), xvi
visualizations, in vignettes, 264

W

waldo package, 195
WARNINGs, R CMD check, 61, 328
website, 281-292
    articles, 259, 268, 289
    deployment, 283
    development mode, 290
    initiating a site, 281-283
    logo, 285
    reference index, 286-287
    vignettes, 259, 288-290
Windows, 30, 39, 41, 44
withr package, 93-94, 141-142, 206-208
withr::defer(), 93, 207
withr::deferred_clear(), 94
withr::deferred_run(), 94, 207
withr::local_options(), 215
withr::local_tempfile(), 217
with_*(), 94

workflows and tooling, 47-62
    check(), 60-62
    creating a package, 47-52
    function documentation, 234-237
    load_all() for test drive, 58-59
    NAMESPACE file generation, 145
    R CMD check, 60-62
    RStudio Project, 52-56
    vignette writing, 260-261
    working directory and filepath, 56
working directory, 56
writing a function, 6-7
writing files during testing, 216
Writing R Extensions, xvii

X

XDG Base Directory Specification, 112

Y

YAML, 262

About the Authors

Hadley Wickham is chief scientist at Posit, winner of the 2019 COPSS award, and a member of the R Foundation. He builds computational and cognitive tools to make data science easier, faster, and more fun, working on packages like the tidyverse for data science and principled software development. He is also a writer, educator, and speaker who promotes the use of R for data science.

Jennifer Bryan is a software engineer at Posit, a member of the R Foundation, and a part of the tidyverse team that maintains more than 150 R packages. Jennifer maintains packages for importing tabular data, working with Google APIs, and simplifying development workflows.

Colophon

The animal on the cover of *R Packages, Second Edition* is a kaka, or nestor parrot (*Nestor meridionalis*), found in native forests of New Zealand. Generally heard before they are seen, kaka are very gregarious and move in large flocks.

Kaka are obligate forest birds that obtain all their food from trees. They are adept fliers, capable of weaving through trunks and branches, and can cover long distances, including over water. They consume seeds, fruit, nectar, sap, honeydew, and tree-dwelling invertebrates.

Although forest clearance has destroyed all but a fraction of the kaka's former habitat, the biggest threat to their survival is introduced mammalian predators, particularly the stoat, but also the brush-tailed possum.

Many of the animals on O'Reilly covers are endangered; all of them are important to the world.

The cover illustration is by Karen Montgomery, based on an antique line engraving from *Wood's Animate Creation*. The cover fonts are Gilroy Semibold and Guardian Sans. The text font is Adobe Minion Pro; the heading font is Adobe Myriad Condensed; and the code font is Dalton Maag's Ubuntu Mono.

CPSIA information can be obtained
at www.ICGtesting.com
Printed in the USA
JSHW050320240623
43688JS00003B/7